LIFE ON THE SEA-SHORE

LIFE ON THE SEA-SHORE

by

A. J. SOUTHWARD, D.Sc.

*Zoologist at the Plymouth Laboratory
of the Marine Biological Association of the United Kingdom*

HARVARD UNIVERSITY PRESS

CAMBRIDGE, MASSACHUSETTS

1965

WINGATE COLLEGE LIBRARY
WINGATE, N. C.

© A. J. Southward 1965
First published 1965

Printed in Great Britain

Preface

THE sea-shore is one of the most restricted of all biological habitats and rarely occupies more than a few metres vertically from low tide to high water mark. Within this small area there is a variation of environmental factors and a richness of life that is not found in less limited habitats. We are very far from understanding the ecology of the shore, and the individual student or field party can still make useful additions to the existing body of knowledge, providing that they organise their work on a sound basis.

Although observation and experiment may be easier on the shore than in many other habitats it is essential that the great variety of animals and plants should be correctly identified. It is assumed in this book that the reader is familiar with the general appearance of all the major invertebrate groups, and can learn to 'run down' the organisms to at least genus or family level, using one of the special guides for the purpose.[1] For maximum benefit some of the commoner animals ought to be identified to species.

It is equally important that tide-levels and other major factors should be correctly measured and that the organisms should be properly sampled. Chapter 8 is devoted to instruction on these and many other practical aspects of the study of the sea-shore. The main part of the book deals with the descriptive and theoretical aspects of life on the shore, starting with the environment and proceeding to an account of the zonation and distribution of the plants and animals. This arrangement is adopted to ensure some degrees of familiarity with the important environmental variables and their effects on the

[1] See Further Reading (p. 140).

organisms; otherwise it is difficult to describe the biological features satisfactorily or to understand the later chapters dealing with adaptation to life on the shore and the causes of zonation. At all stages emphasis is placed upon active participation in field and experimental work.

I would like to express my thanks to my wife, Dr. Eve Southward, for encouragement, discussion and active help in preparing this book, and to Mr. W. H. Dowdeswell, the Editor of the series, and Mr. J. S. Colman for helpful and detailed criticism of the manuscript.

Plymouth, March 1965   A. J. S.

Contents

CHAPTER		PAGE
1	Introduction	1
2	The Environment	3
3	Rocky Shores	15
4	Sandy and Muddy Shores	49
5	Estuaries and Lagoons	74
6	Adaptations to Life on the Shore	86
7	The Causes of Zonation	114
8	Methods of Studying the Sea-shore	121
9	Further Reading	140
	Glossary	147
	Index	149

List of Plates

PLATE

I	A rocky shore in the south-west	*facing page* 22
II	Three common brown seaweeds	23
III	Some animals from the upper part of a rocky shore	54
IV	More high-level animals and plants	55
V	Plants and animals from below mid-tide level	86
VI	Plants exposed at low spring tides	87
VII	Two beaches of coarse sand	118
VIII	A muddy shore in a large estuary and a mud shore showing head shafts to the burrows of *Arenicola*	119

Table 1

Definition of intertidal zone on rocky shores and zones of benthic marine animals

Description	Zone (fine)	Zone (broad)
Extreme limit of spray or salt air		SUPRALITTORAL ZONE
Maritime land lichens and salt-tolerant higher plants		SUPRALITTORAL ZONE
Extreme limit washed by tides or waves	SUPRALITTORAL FRINGE or LITTORAL FRINGE	
Marine lichens, littorinid molluscs, isopod crustaceans	SUPRALITTORAL FRINGE or LITTORAL FRINGE	INTERTIDAL ZONE
Level reached by the higher tides and waves	MIDLITTORAL ZONE	INTERTIDAL ZONE
Barnacles, limpets, green algae, smaller brown algae, some red algae	MIDLITTORAL ZONE	INTERTIDAL ZONE
Level exposed to air by lower tides only	INFRALITTORAL FRINGE or SUBLITTORAL FRINGE	INTERTIDAL ZONE
Large brown algae, many red algae, sometimes calcareous green algae; sometimes also marine flowering plants	INFRALITTORAL FRINGE or SUBLITTORAL FRINGE	INTERTIDAL ZONE
Extreme level exposed to air by tides or waves		SUBLITTORAL or INFRALITTORAL ZONE
Some large brown algae, red algae, marine flowering plants		SUBLITTORAL or INFRALITTORAL ZONE
Extreme level at which there is insufficient light for the plants to grow		SUBLITTORAL or INFRALITTORAL ZONE

1

Introduction

THE sea-shore is best defined as that part of the coastline extending from the lowest level uncovered by the tides up to the highest point washed or splashed by the waves at the highest tides. This region, which corresponds only in part with the legal term foreshore, is more correctly termed the intertidal zone or the littoral zone (from the Latin *litus*, meaning shore) (Table 1). The lower limit of this intertidal zone is rather artificial, for much the same plants and animals may be found in the sublittoral zone, which is usually taken to extend down from the low tide line as far as there is sufficient light for plants to grow (as much as 100 metres deep in clear water). Above the intertidal zone marine life ends abruptly. The supralittoral zone lies beyond the reach of the waves, though subjected to salt spray carried by the wind, and bears only a sparse fauna and flora of salt and drought-resistant forms of terrestrial origin. The relationships of all these zones are shown in Table 1.

The intertidal zone is astonishingly rich in both numbers and species of animals, and all groups other than amphibians, reptiles, birds and mammals have resident representatives on the shore.[1] For this reason it is one of the best training grounds for the aspiring zoologist, as recognised for the past eighty years or more. Plant life is less well represented, though the absence of mosses, ferns and most angiosperms is compensated for by the abundant species of algae. Of course, not all the animals and plants can be expected to be found in one place. The greatest diversity occurs on rocky shores fringing the great oceans; more and different animals will be found burrowing

[1] The sea-otters of the north Pacific and the lizards of the Galapagos Islands are not truly intertidal.

2   INTRODUCTION

into the sandy and muddy shores of protected coves and harbours, though here only a few small and ephemeral plants can exist; and the least life occurs on gravel or shingle beaches where little may exist apart from microscopic organisms deep down under the beach in the film of water between the particles of gravel. In many places along the western coasts of Britain all types of shore exist within a few miles of each other, as for example in the Isle of Man, South Wales and South Devon, where field centres and similar facilities exist, and in Scotland and Ireland.

It is impossible in the present volume to discuss the biology of the various marine and coastal animals that merely feed on the sea-shore at times, notably the birds and off-shore fishes. Information on these will be found in some of the references given at the end of the book in the section on Further Reading.

In the following chapters, when an animal or plant is first mentioned its common name is given first—if it has one—then its full scientific name. The scientific names follow those used in the *Plymouth Marine Fauna* and in Dr. M. Parke's check-list of marine algae (see Further Reading).

The engraving below, taken from a Victorian work on Natural History, shows the rocky and sandy shore at Kynance Cove, Lizard, Cornwall.

2

The Environment

THE sea-shore is essentially an extension of the marine province, and the animals and plants that live there are almost all of directly marine origin.[1] Most of their activities are reduced when they are temporarily exposed to the air by tides or waves and are resumed in full only when they are once more covered by the sea, which is their natural medium for gaseous exchange, photo-synthesis, feeding and breeding.

Sea water is a solution of various salts, most of them fully ionised, with a strength varying from 30 to 35 parts per thousand on most shores, sometimes rising to 40‰ in certain circumstances. The commoner constituents of sea water are listed in Table 2, and a formula for artificial sea water is given in Chapter 8. The bicarbonate content is most important, as it has a substantial buffering action, the normal range of pH in the sea being very small, from 8·2 to 8·4. Water held between particles in sandy beaches may become more acid, though the pH rarely falls below 7, while in pools and sheltered inlets on calm days the photosynthetic activity of seaweeds may remove sufficient bicarbonate to raise the level to pH 9 or more.

Owing to the great specific heat of water and the vast extent of the oceans, the sea acts as a source of heat and as a heat sink. It gives up some of its store of heat in the winter (some derived by flow from warmer ocean waters) and takes up heat in summer from the rocks and hot sand. Thus, although the animals and plants of the shore are subjected to a greater daily and seasonal variation than those in the sublittoral zone, they may experience less severe changes than do terrestrial

[1] Insects and a few other arthropods of the intertidal zone are of terrestrial origin.

Table 2

Composition of sea water, from recent analyses

ion	Content in g/kg.
Na^+	10·556
K^+	0·380
Mg^{++}	1·272
Ca^{++}	0·400
Sr^{++}	0·0085
$H_3BO_3^-$	0·026
Cl^-	18·980
SO_4^{--}	2·649
HCO_3^-	0·140
Br^-	0·065
F^-	0·004
SiO_3^{--}	trace
NH_4^{++}	,,
NO_2^-	,,
NO_3^-	,,
PO_4^{---}	,,
Fe^{++}	,,
Mn^{++}	,,
Cu^{++}	,,
Zn^{++}	,,
Mb^{++}	,,
Va^{++}	,,
Co^{++}	,,

And approx. 5·4 ccs. O_2 per litre at 20° C. 8·08 ccs. per litre at 0° C.

species (Fig. 1). However, the daily changes are very abrupt, while those on land are more gradual.

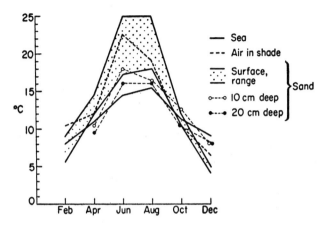

FIG. 1. Temperatures observed on a beach at bi-monthly intervals during one year. Comparison of sea temperatures close to the shore, shade air temperature near the beach, and the range of surface temperature found at various points on the beach with the temperatures found at depths of 10 cm. and 20 cm. in the sand. The high sand-surface temperature, sometimes higher than shade air temperature, is due to direct heating effects of sunlight.

Tides

Except in a few regions of the world (e.g. the Mediterranean and Baltic Seas) the tide is the most important single factor governing life on the shore. The periodical rise and fall that we call tide is primarily caused by attractive forces exerted on the earth by the sun and moon. The forces vary with the relative positions of the three bodies and the rotation of the earth, and are modified by the effects of friction and changing depth of water in different parts of the world, and by the natural periods of oscillation of the ocean basins and coastal inlets, hence the phase of the tide and the height of its rise and fall vary greatly. The tidal period also varies; ideally it would be expected that a rise and fall would occur twice in each 24 hours (semi-diurnal tide) near the equator, but only once (diurnal tide) near the poles, depending on the declinations of the sun and moon. In practice this does not happen, and examples of both types of tide can be found in tropical and polar areas. A mixture of the two (mixed tide), is also found, for example in large areas of the Pacific, while in a few shallow-water places more frequent changes may be experienced (e.g. quarter-diurnal tides). The latter category includes the famous 'double tides' of the part of the south coast of England between Portland and Portsmouth, but elsewhere in the British Isles the tide is almost purely semi-diurnal.

Since it is possible to predict the relative movements of the earth, moon and sun, it follows that it is possible to predict the tide-generating forces. Tide tables can thus be calculated for any place where previous observations of tidal heights and times have been made over a period of at least some months. The published tide tables require intelligent use, and it is obviously best to obtain those that give predictions for low water as well as for high water, and for a place as near as possible to that being visited (see Further Reading). Approximate 'corrections' to the tide tables can be made for a shore not listed, by as few as 24 hours' consecutive hourly readings of the height of the tide there compared with simultaneous

---

FIG. 2. Examples of tide curves as might be drawn either by a recording tide gauge or by a tide-predicting machine: (1) a semi-diurnal tide, as on most parts of the English coast, showing how the range between the twice daily high and low waters decreases on passing from spring tides to neap tides; (2) a mixed tide, predominantly semi-diurnal, but showing increasing differences between the two daily low waters on passing from spring tides to neap tides (from Singapore); (3) a mixed tide in which the diurnal component is strongest, and the semi-diurnal tide appears only when the diurnal tides are weak, at the equatorial tides (Gulf of Mexico); (4) an almost fully diurnal tide, with faint appearance of a semi-diurnal component on the equatorial tides (Gulf of Mexico).

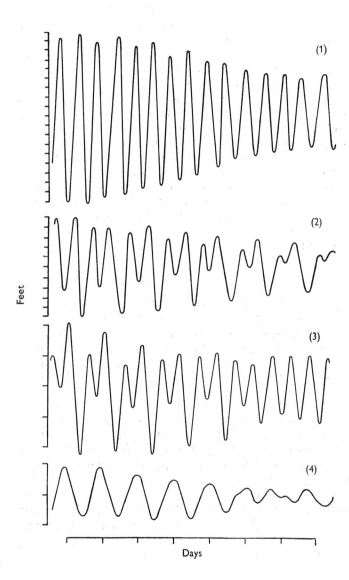

readings at a place for which tables are already available (see Chapter 8 for full method).

When venturing onto an unfamiliar shore some allowance must be made for unpredicted changes in sea-level, due to non-tidal factors. For example, the sea may not fall as far as expected when the wind blows on shore or when the barometer is low, whereas a high barometer or an off-shore wind may cause the level to fall lower than expected. Such effects are infrequent on British shores, where less than 5% of predictions differ by as much as 0·3 m. from the actual level, and greater differences are extremely rare. However, the particularly impressive ones known as storm surges have long-lasting effects on the shore line. In the Mediterranean the true tides are small (a few decimetres only) and the irregular changes in sea-level due to winds or barometric pressure are then noticeable.

With semi-diurnal tides the rise and fall is greater when the sun and moon are in the same phase (spring[1] tides) and least when the attractive forces of the two bodies are at right angles relative to the earth (neap tides). The interval between two sets of spring tides is about a fortnight, and the spring/neap differences reach a maximum twice a year, near the vernal and autumnal equinoxes. Where the diurnal tide predominates the corresponding terms for the greater and lesser tides are 'tropical' and 'equatorial'; being connected with the declination of the moon.

In most places the rise and fall of the tide corresponds to a cosine curve, and changes in level take place more slowly around the times of high tide and low tide, and ebb and flow being most rapid half-way between. Some examples of tide curves, as traced by an automatic tide-registering gauge or by a tide-predicting machine are shown in Fig. 2. Table 3 gives some of the technical terms often quoted in connection with tides, which are useful when referring to the positions of animals and plants on the shore; the commonly accepted

[1] The term spring here has no connection with season, but is derived from the Old English *springan*, meaning to rise.

Table 3

Tidal abbreviations and meanings

Abbreviation	Full name	Definition
EHWS	Extreme high-water springs	The highest level reached by the sea on the greatest tide or tides of the year (sometimes 'high water equinoctial spring tide').
MHWS	Mean high-water springs	The average of the higher levels reached by the sea on each fortnightly set of tides.
MHWN	Mean high-water neaps	The average of the lowest high waters reached by the tide on each fortnightly set of tides
E(L)HWN	Extreme (lowest) high-water neaps	The lowest of the recorded high tides of neap tides. Below this level the shore is wetted on every tide.
MTL	Mean tide-level	The average of a long series of observations on the heights of high tide and low tide.
MSL	Mean sea-level	The average level of the sea calculated from a long series of observations on the height of the tide at regular, short, intervals (e.g., hourly intervals for one month or more). [MTL and MSL are not quite the same in value as the tidal curve is seldom exactly harmonic.]
OD	Ordnance datum	Approximately mean sea-level, the level to which all land surveys are related. There are two in use in Britain differing by a fraction of a foot, OD (Newlyn) and OD (Liverpool), and it is important to know which is employed on the map or survey being used.
E(H)LWN	Extreme (highest) low-water neaps	The highest recorded low water of the neap tides. Below this level the shore is not exposed to the air on every tide of the year.
MLWN	Mean low-water neaps	The averages of the higher low tides of each fortnightly set.
MLWS	Mean low-water springs	The average of the lowest levels reached by the tide on each fortnightly set of tides.
ELWS	Extreme low-water springs	The lowest level to which the tide falls on the greatest tides of the year (sometimes 'low-water equinoctial spring tide').
CD	Chart datum	A level, sometimes near MLWS, sometimes ELWS, to which the soundings on a naval chart are referred. Generally defined as a 'level below which the tide seldom falls'.

Salinity

Most animals and plants that live on the sea-shore can tolerate a much wider range of salinity of the water than can those that live in the sublittoral zone or in the ocean, and the variations due to rainfall on the shore or to local drainage from the land usually have no effect. However, the situation is very different in estuaries, where the volume of fresh water coming down is large, and many estuarine organisms show special adaptations. This matter is dealt with in Chapter 5.

Waves

The action of the waves affects life on the shore in two ways. First there is a supplement to tidal effects since frequent wetting or splashing above the height of the tide at any point may allow animals and plants to live higher up on the shore than they would in shelter. This aspect is described more fully in the chapter on rocky shores. Secondly, there is a mechanical effect, composed of pounding and tearing forces. On a rocky shore that is constantly wave-beaten the organisms are very firmly attached to the rock, most of the animals having the simple hemispherical shape of a limpet or barnacle, and the plants an unbranched, or only slightly branched, short and sturdy thallus.

Wave action is of supreme importance on shores of loose material. It influences the size of particles that remain on the beach (the more the wave action the larger the particles) and hence controls the type of beach, whether muddy, sandy or gravelly. It affects the beach profile, particularly at high- and low-water marks where the pounding action is felt for longer than at intermediate levels (see p. 7), and this effect, together with the size of the particles, controls the amount of water held in the beach during low tide. All these factors act in concert to govern the types of animals burrowing in the beach, and their abundance.

Wave action is difficult to measure. Although dynamometers have been used to assess mechanical shock of individual waves, sometimes reaching 6,000 pounds per square foot, it is the frequency of wave action on a shore that is more important to the plants and animals. A rough estimate of likely frequency of wave action can be made by observing which winds and wind forces cause appreciable wave action (measured in relative height against fixed structures such as piers and breakwaters), and then noting from meteorological records how often such winds are experienced. Another estimate of the same type can be made by observing the angle over which a beach is exposed to the open sea for more than some specified distance, allowance being made for prevailing winds. However, all such estimates are rendered inaccurate by the occurrence of swells, which have no relation to local wind, and by the ease with which waves are refracted around headlands. Thus in most cases we can say only that one beach experiences more wave action than another, and not by how much the two beaches differ. This is unfortunate, since (as will be noted later) there are certain characteristic changes in the plants and animals that can be correlated with variations in the frequency of wave action.

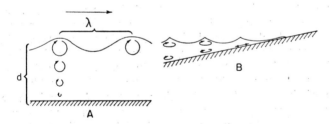

FIG. 3. A, showing the circular movement of particles under the influence of wave motion, and how the influence decreases with depth (d). Wave-length (λ) is the distance between crests.
B, how the circular motion of particles is transformed to an up-and-down-the-beach motion as shallow waves impinge on a gently sloping shore.

Observation shows that there are two principal types of wave on the shore. The less destructive type (sometimes termed constructive) is of fairly long wave-length (distance between crests) and as it reaches the shallows it breaks gently into a forward rush of water up the shore (wash) without much spray being thrown up. The diagram (Fig. 3) shows how the normal circular or near circular movement of water particles in deeper water is transformed into a movement up and down the beach. The more destructive type of wave is the plunging type, in which the foot of the wave is retarded as the water shallows, causing a towering crest to develop and curl over before breaking. This type of wave is generally of shorter wave-length and produces less wash and more splash than the other type. It is responsible for much of the erosion of beach materials that occurs on sandy and gravelly shores (see Fig. 4).

Substratum

The type of shore dictates the types of plant and animal life. A solid rocky shore presents a stable environment where many long lived species can survive and supports the greatest diversity of species. On such shores most of the animals are sessile or semi-sessile, typically sponges, anemones, barnacles and limpets. Beaches of loose material, however, are much less stable, and support less diverse forms of life; a constantly shifting bank of shingle is a virtual biological desert. The greatest extent of burrowing life, chiefly worms and molluscs but including crustaceans and echinoderms, is found on the relatively more stable beaches of fine sand or muddy sand. Entirely muddy shores, though more stable than sandy shores, are apparently less suitable for many burrowing animals because the fine particles clog the delicate gills and feeding mechanisms of filter feeders.

The various types of shores are very unevenly distributed. The typical western coasts of Britain consist of hard rocks—granites, schists and the older slates and limestones—which are extremely resistant to erosion by the waves. The irregular nature of the coastline is due in part to local variations in hardness and hence of resistance to the sea, but also to a rise in sea-level which has flooded former valleys excavated by glacial and aerial erosion. With such resistant rocks extensive shingle beaches are rare, since shingle is nearly always derived from the cliffs by erosion and subsequent re-working on the shore. However, along parts of the Outer Hebrides and the western coast of Ireland coarse sandy beaches are locally abundant.

The sandstones, chalks and limestones found further east in Britain are much more liable to leaching and erosion by the waves, and deposits of sand and shingle become more frequent.

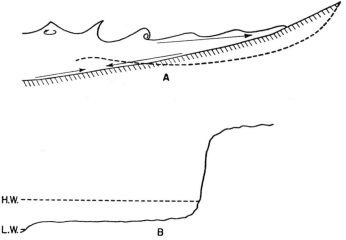

FIG. 4. A, the effect of plunging (destructive) waves on the beach profile. The solid line and hatching show the profile under gentle wave action, the broken line the profile resulting from plunging waves. The solid arrows show water movement, and illustrate how particles moved from higher up the beach accumulate below low tide to form a bar.
B, a cliff and rocky beach in cross-section, showing a 'wave-cut platform' above low water.

Along most of the south-eastern coast where harder rocks are entirely absent, the beaches consist entirely of sand and gravel; the few cliffs of recent geological age, such as sands and clays, are being very rapidly eroded.

Wave action is most effective in removing material from the beach and cliffs at high water, particularly when meteorological conditions result in a higher tide than usual (e.g. storm surges). Generally speaking, if the cliffs do not fall sheer to great depths and if eroded material is not removed by water currents, an extensive 'wave cut platform' is formed just above the level of low water (Fig. 4) by the cutting back of the cliff. Eventually the effects of wave action may be reduced since some of the energy is dissipated in passing over the platform.

If we omit the shingle beaches, which are largely devoid of life, the remaining types of shore can be grouped into two categories: the solid rocky ('eroding') shores, including those boulder beaches that can support life, and the 'depositing' sandy and muddy shores. These two groups offer such different environments and have such a different fauna and flora that they must be considered separately.

3

Rocky Shores

THERE are several special factors which influence the animals and plants living on rocky shores. The type of rock itself has some effect. Certain animals cannot survive the desiccation that can take place through their attachment or base on porous rocks such as chalk, and at the other extreme, very hard igneous rocks may have too smooth a surface for a plant or animal to attach itself firmly enough to resist the force of the waves. However, more important aspects of the general biological appearance of a rocky shore are attributable to the angle of the slope presented to the sea, whether it is smooth or broken up into a series of reefs or platforms, and its orientation to the sun. Many plants do best on broken reefs in shade from the sun, while the greatest variety of fauna is found under overhanging ledges.

An introduction to the type of life found on rocky shores is best provided by the slate rocks of a partly sheltered bay on one of the western coasts. After considering the general features of a shore of this type we can go on to note the differences found on other shores and in other places.

A typical rocky shore (See Plate I)

On this shore as on many others of the same type, the way from the road lies through a small valley or combe down which a small stream flows. The stream fans out as it reaches a ridge of small boulders and pebbles that marks the upper limit of wave penetration at high tide during stormy weather. Just below this ridge of pebbles, which is formed by the almost explosive force of the breaking waves, we cross a bank of rotting seaweed, left behind by the waves. The seaweed is particularly in evidence after a gale, and consists of the fronds and stipes of the large brown oarweeds, or laminarians. These have been torn from their attachment lower down the shore or in the sublittoral zone and, now washed up at high water, provide food for many scavenging animals. For those who like grubbing, searches in the weed will reveal many amphipods of the sand-hopper type, such as *Orchestia* and perhaps *Talitrus* (Fig. 26), if there is some sand present. Later in the season

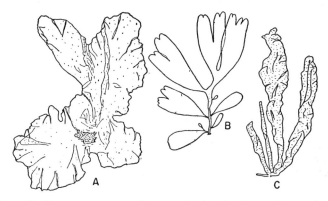

FIG. 5. Some common small seaweeds of rocky shores. A, *Porphyra umbilicalis* (red) from HWN; B, *Rhodymenia palmata* (red), MTL and below; C, *Enteromorpha intestinalis* (green), up to HWS.

grubs and adults of Diptera will be more abundant, especially the beach fly *Coelopa*, which is a well-known plague of the popular shingle-beach resorts.

As we walk away from the stream towards either side of the bay the gravel and boulders give way to solid rock and the first zones of algae are encountered. Where the rocks are moist, due to fresh water seepage or other reasons, the green alga *Enteromorpha* will be obvious. This form and related green seaweeds occur further down the shore also, but are not so noticeable there, as they are grazed down by limpets and other herbivorous gastropods. Proceeding along high-water mark, away from the moist patches, the uppermost brown alga is channelled-wrack, *Pelvetia canaliculata* (Plate IV), a small slightly branched plant which appears brown when wet but soon dries out to a dark brown or blackish tuft. A short distance below the *Pelvetia*, possibly overlapping with it to some extent, is a second species of brown alga, flat-wrack, *Fucus spiralis* (Plate II). These two occur in narrow belts above the range of neap tides, and may be accompanied by scattered barnacles and periwinkles (see p. 31).

As the tide drops below the level of high water of neap tides more shore animals can be seen, with further species of algae. bladder-wrack, *Fucus vesiculosus* (Plate II), with air bladders in its fronds, grows at this level as well as further down the shore; in its shade are barnacles and limpets on the rock beneath. Other more snail-like gastropods (Fig. 8) feed on the fronds of the weed, notably the purple-striped top-shell *Gibbula* and the yellow or brown periwinkle *Littorina littoralis*, also the white-shelled dog-whelk *Nucella*, which feeds on the barnacles and on mussels. In shallow pools at this level we may be able to find several other animals, including small prawns (*Leander squilla*), the red form of the common sea anemone *Actinia* (Plate III, Fig. 6), and in pool crevices the waving tentacles of the snakelocks anemone, *Anemonia sulcata*, sometimes green, sometimes brown, but always tipped with mauve. Below neap tide level patches of the knotted-wrack, *Ascophyllum nodosum* (Plate II) begin, and half-way down the shore this species takes over from *Fucus vesiculosus*. The fronds are olive green, with brighter yellow patches where gametes are produced, and may be a metre or more in length. The largest fronds of *Ascophyllum* are the product of many seasons' growth, and during this time other animals and plants may settle on them and grow. There are purple-red tufts of the red alga *Polysiphonia*, and various filamentous brown algae, such as *Pilayella;* pink colonies of the naked hydroid *Clava squamata* (Fig. 10) are obvious as the tide leaves the seaweed but will be very much dried up by the time the tide has fallen to low water; there are also patches of the sawtoothed stems of the thecate hydroid *Dynamena pumila*. There is less to be found on the

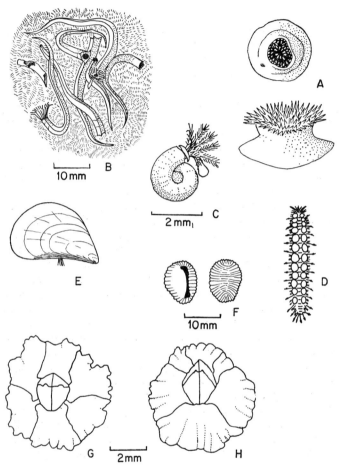

FIG. 6. Some common rocky-shore animals. A, the sea-anemone, *Actinia equina*, seen from above while out of water, and from the side while under water, found in pools and wet places from HWN down; B, the tube-worm, *Pomatoceros triqueter*, found in groups at LWN and below; C, the tube-worm, *Spirorbis borealis*, found mainly on *Fucus*, other species occurring on rocks and on other seaweeds; D, the scale-worm, *Lepidonotus clava*, found under

rock under the fronds of the *Ascophyllum*, and probably only a few large limpets will be seen. The pools, however, have a more interesting fauna and flora, with fish, young hermit crabs (Fig. 7) living in empty shells of *Gibbula*, *Littorina* and *Nucella*, and many species of red algae. On this particular shore the *Ascophyllum* stops short of low water of neap tides, and a further species of *Fucus* becomes dominant. This, serrated-wrack, *Fucus serratus* (Plate V), has very smooth, flattened, slippery fronds with serrated edges, and is much less subject to overgrowths of plants or animals than *Ascophyllum*. It is generally more sparse and covers the ground less closely, so that there may be quite an undergrowth of short red algae, including *Rhodymenia* and *Laurencia* (Fig. 5 and Plate V), and possibly the encrusting pink growths of *Lithothamnion* and the jointed calcareous stems of the equally pink *Corallina*. These two genera of algae are rarely found on the open rock above this level, but grow in pools up to high water of neap tides. In places there may be a few limpets or barnacles, but these are obviously fighting for space on the rock with the encrusting algae and with the calcareous tubes of *Spirorbis*, a small serpulid polychaete (Fig. 6).

There is a dramatic change in the appearance of the shore as the tide falls to low-water spring level, and the large oarweeds or laminarians are exposed. These may be one to two metres long, including the branched holdfast clinging to the rock and the stout stipe bearing a broad, perhaps tattered, thallus (Fig. 13; Plate VI) *Laminaria digitata* is the commonest and the thinner fronds of *L. saccharina* are easily distinguished, while the twisted stipes of *Saccorhiza* may be discerned in places, Far out, at the visible edge of the Laminarian zone are

---

stones from LWN down; E, the common mussel, *Mytilus edulis* of which small forms are found from MTL down; F, the small cowrie-shell, *Trivia arctica*, found at LWS in shaded places; G and H, the two common acorn barnacles of Europe, *Chthamalus stellatus* (G) and *Balanus balanoides* (H). A, D and E, natural size; others enlarged.

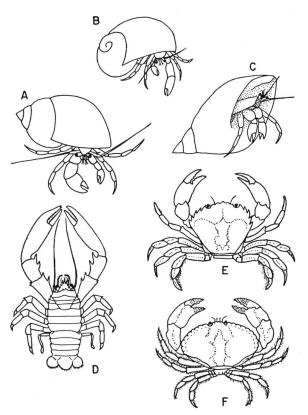

FIG. 7. More rocky-shore animals. A, the common hermit crab, *Eupagurus bernhardus*, juvenile form frequent in pools about LWN (living in shell of *Littorina littorea* in this case); B, for comparison, *Diogenes pugilator*, found at LWS on some exposed sandy shores; C, a hermit crab of warmer waters, *Clibanarius misanthropus*, found in the near MTL or higher-level rock pools in S.W. England. Note that A has its large claw on its right hand, B is left handed, and C has almost equal claws, hairy and striped with blue. D, the common squat-lobster, *Galathea squamifera*, found under stones at LWS; E, the common shore crab, *Carcinus maenas*, under stones and in

the tall, stout stipes of *L. hyperborea* sticking out of the water, the most strongly perennial species of this group.

The laminarians occur in a dense stand, canopy or forest, and little else but *Lithothamnion* can grow on the rock beneath them. The stipes of *Laminaria hyperborea* support many epiphytic seaweeds, and the fronds of all species are the home of the pretty little gastropod *Patina*, which has three iridescent blue-green stripes on its otherwise transparent shell. When the animal grows to maturity it becomes just another dull limpet-like form, and is to be found inside the holdfasts of the larger laminarians. These holdfasts have a very rich fauna in their crevices and hollows, including many young crabs, other crustaceans and many worms, of which the small ragworm, *Nereis pelagica* (Fig. 9) may be the largest. The pools at this low level may not be as rich in species as those higher up the shore, unless they are in deep gullies where the laminarians do not grow. In such places, and under overhanging faces, there is an infinite variety of encrusting and mobile life. White, yellow, green, orange, brown and grey sponges and red and brown parches of hydroids may be seen; the common anemone *Actinia* is present in profusion here, including green and brown forms and a spotted strawberry variety, as well as the usual red type. The tiny anemone *Sagartia elegans* occurs in a variety of patterns based on orange, white and brown, more rarely pink. The little cowrie shell *Trivia* (Fig. 6), a poor relation of the large tropical forms whose shells have been used as currency, occurs here and there, and feeds on the colonies of tunicates, some white and amorphous, some with patterns of red and green and gold, and sometimes violet (*Botryllus*, *Botrylloides*, Fig. 11). The tube worms *Pomatoceros* and *Spirorbis* (Fig. 6) are also found in such places, as well as under large stones. In the latter habitat, in addition to the encrusting forms already

---

pools from MTL down; F, juvenile edible crab, *Cancer pagurus*, found under stones at LWS. A to D approximately natural size, E and F reduced.

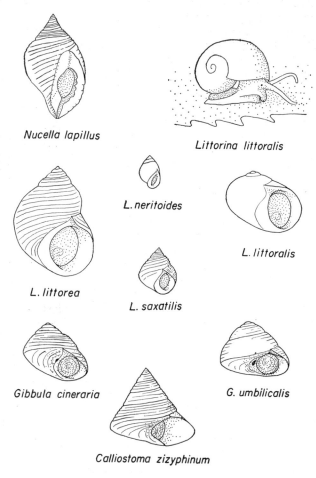

FIG. 8. Common gastropod molluscs of the shore, natural size.

PLATE I. A rocky shore in the south-west. The reef in the foreground is partly sheltered from wave-action, as the abundance of knotted-wrack (*Ascophyllum*) suggests. The tide pools and crevices have a rich and varied fauna.

PLATE II. Three common brown seaweeds: *top left*, flat-wrack, *Fucus spiralis* from near HWS; *top right*, bladder-wrack, *Fucus vesiculosus*, from about HWN; *bottom*, knotted-wrack, *Ascophyllum nodosum*, from MTL.

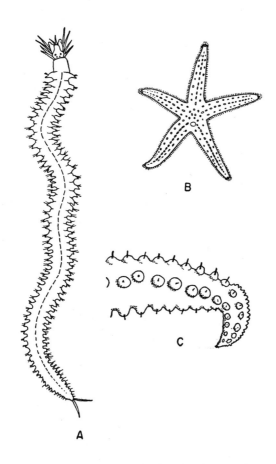

FIG. 9. Further rocky-shore animals: A, the small rag-worm, *Nereis pelagica*, found in holdfasts ('roots') of laminarian seaweeds; B, small example of the common starfish, *Asterias rubens*, from under stones LWN down; C, one arm of the glacial starfish, *Marthasterias glacialis*, sometimes found at LWS, showing differences from *Asterias*. A enlarged, B and C natural size.

mentioned, we find a selection of chiefly sublittoral animals which are forced to conceal themselves at low tide. The blue and red squat-lobster *Galathea* (Fig. 7) is more frequent than young specimens of the edible crab (Fig. 7); the reddish common starfish *Asterias* (Fig. 9) may outnumber the more roughly spotted, greenish, *Marthasterias*, while the globular

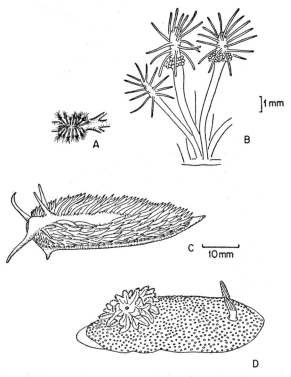

FIG. 10. Rocky-shore animals: A, the hydroid, *Clava squamata*, found on knotted wrack (*Ascophyllum nodosum*), from MTL down; B, part of A enlarged; C, a nudibranch mollusc, *Aeolidia papillosa*, that feeds on hydroids and anemones; D, another nudibranch, *Archidoris britannica*, often found on sponges and ascidians at LWS.

sea-urchins *Echinus esculentus* and *Psammechinus miliaris* must be searched for diligently. Also present under stones are variously coloured brittle-stars, *Amphipholis* and *Ophiothrix*, together with some of the larger shore fishes: gobies, blennies (Fig. 12) and sea-scorpions, as well as occasional barb-nosed rocklings and young conger eels. Another fish that may be found is the common pipe fish, famous for having the male carry the eggs around until they hatch. The various sucker-fish or 'sea-snails' attach themselves firmly to stones by their modified ventral fins.

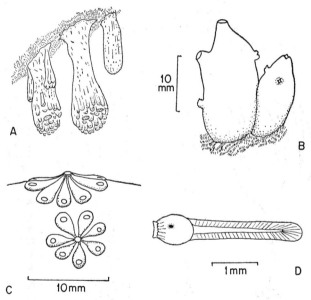

FIG. 11. Some ascidians of rocky shores: A, *Morchellium argus*, a colonial form from under ledges at LWS; B, the non-colonial, but gregarious form, *Dendrodoa grossularia*, from under stones, MLWN and below; C, part of a colony of *Botryllus schlosseri* showing the stellate groups round a common cloacal aperture; D, tadpole larva of *Dendrodoa* (see p. 102).

Rock goby, *Gobius paganellus*

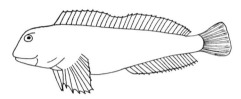

Shanny, *Blennius pholis*

FIG. 12. Two common shore fishes. Natural size.

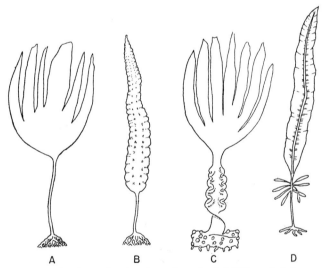

FIG. 13. The large laminarians found at LWS and below: A, *Laminaria digitata* (*L. hyperborea* looks very much the same reduced to this size); B, *L. saccharina*; C, *Saccorhiza polyschides*; D, *Alaria esculenta*. About 1/20th natural size.

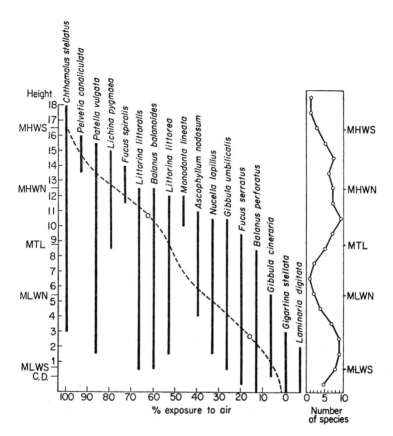

FIG. 14. Distribution of some plants and animals in relation to tide level, showing the range of each species, not its abundance. The broken line indicates the % of time the shore is exposed to the air at each level. The subsidiary graph at the side shows the number of species (including some not shown on the main part of the figure) that have an upper or lower limit at each foot vertically. It can be seen that a maximum number of upper or lower limits of species is found at two levels, below MHWN and below MLWN. These are levels where there is a change in slope of the exposure curve, as denoted by the open circles. (Wembury, S. Devon, after Colman.)

Shelter

The shore described above and featured in Fig. 14 is subject to a moderate amount of wave action, and although the undersides of stones may have a light coating of silt there is still no real deposit of mud on the shore. With increasing shelter from the waves more and more mud is encountered and the number of organisms becomes very much smaller. Many of the red algae soon disappear, followed by the laminarians, so that the part of the shore below low-water neap level may bear only a short turf of the hardier red algae. In extreme shelter the fucoids tend to cover most of the shore from just about MHWN (see Table 3) down to about mid-tide level and below, *Ascophyllum* being dominant, while barnacles and limpets are very scarce. At low water there may be only a few species of encrusting sponges, together with mussels and oysters. Where much mud is held in suspension in the water little light can penetrate, and the few algae then present are found mainly near high-water mark. However, the effect of mud in suspension is experienced mainly in estuaries, where the salinity is lower, and hence it is difficult to separate the effects of the two factors (see Chapter 5).

In passing from a sheltered place to a wave-beaten point we can see the influence of wave action on the upper limits of the animals and the plants (cf. Fig. 16). There is a lifting of the belts or zones, relative to low-water line, so that in a very wave-beaten place the average tide levels have very little reference to the levels of the plants and animals. We are then forced to refer to 'physiological tide levels', which assume that an organism is found at an equivalent level of immersion, whether due to tides or waves or both (see p. 10).

Wave action

As we leave the sheltered bay described earlier and walk out towards an adjoining headland the brown fucoids become much scarcer, first *Ascophyllum* and then *Pelvetia* dropping out, while the normal forms of *Fucus spiralis* and *F. vesiculosus* are replaced by dwarf or vesicle-less forms which can better resist the force of the waves (cf. Fig. 15). There is a conspicuous boundary line on the upper shore, marking a transition from a yellow, or whitish-coloured belt below to a dark or even black band above (Plate IV). On close inspection this boundary

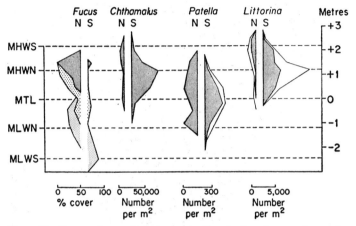

FIG. 15. Comparison of the zonation and abundance of certain plants and animals at nearby wave-beaten (S) and partly sheltered sites (N). (South and north sides of the Breakwater in Plymouth Sound, after Southward and Orton.) Three species of Fucus are shown in succession downwards: *F. spiralis*, *F. vesiculosus* and *F. serratus*. For *Patella* the proportion of *P. vulgata* (grey) to two other species (white) is shown. *Littorina* includes *L. neritoides* (grey) and *L. saxatilis* (white).

is seen to be the upper limit of the barnacles, which are so abundant and crowded together that the rock may be completely hidden.

Several species of barnacles may be present on British shores, but only two contribute to the 'barnacle line' (Fig. 6). In the south and west this level is occupied by *Chthamalus stellatus* which occurs up to and above MHWS. It is a southern form

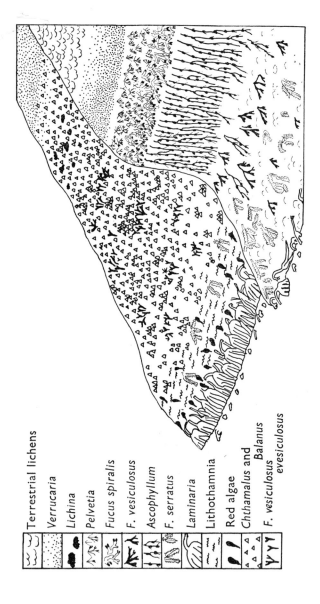

Fig. 16. Pictorial representation of the zonation of the main species of plants and animals on wave-beaten rocks (back) and very sheltered rocks (front). (Anglesey, N. Wales, after Lewis.)

which can stand great heat and desiccation. In the eastern English Channel, up the east coast as far as Wick and in parts of the Irish Sea and Scottish lochs the upper barnacle is *Balanus balanoides*, a northern form which is less resistant to heat and drought: the barnacle line then lies at about MHWN or a foot or two higher. In some areas both species may be present at high water, the *Chthamalus* forming an upper sub-zone from HWS to HWN; in other areas *Balanus balanoides* is found at MTL and below, where other species of barnacles also occur (e.g. *B. perforatus* and *B. crenatus*). The black band above the barnacles is produced by patches of encrusting lichens, particularly *Verrucaria maura*, which supplies most of the black colour, and tufts of the lichen *Lichina confinis*, which are also black. Another, larger species of lichen, *L. pygmaea* occurs in the upper part of the barnacle zone where algae are not abundant (Plate IV). The only abundant animals above the barnacle zone are the periwinkles, *Littorina saxatilis* and *L. neritoides* (Fig. 8), present in thousands or more per square metre. A few barnacles may straggle up this far in crevices, but not enough to influence the colour of the zone. The littorinids are not confined to this upper band but occur in equally large numbers among the barnacles, between them and inside their empty shells, as far down as mid-tide. The upper boundary of the lichen and littorinid zone is not very clearly marked and the downward extension of growths of maritime land lichens and clumps of flowering plants (e.g. sea thrift, *Armeria*) is very variable. Nevertheless, they mark the true limit of marine forms.

Looking down the shore from this level we see that the barnacles, which are accompanied by large numbers of small limpets, continue a long way down the shore, to a level approximately equivalent to that of *Fucus serratus* in shelter. If the shore is not too wave-beaten several species of brown algae may be found between the lower limit of barnacles and the start of the laminarian zone. Among these the thongweed *Himanthalia elongata* (Plate V) is most distinctive. It may occur mixed with *F. serratus*, or (in a few places in the south and west) may give way to a belt of *Bifurcaria bifurcata*. This fucoid and *Cystoseira* are present in pools from high-water mark down, but are found out of the water only on very low spring tides. The *Cystoseira* species are more common in subtropical regions, and are not easy to identify, though some display a distinctive blue-green fluorescence of the growing young shoots under water.

Where wave action is very strong *Laminaria* may be unable to exist and its level on the shore may have a sparser growth of *Alaria esculenta*, (Fig. 13) or of turf-like red algae, together with luxuriant growths of *Lithothamnion*. The tide pools of a wave-beaten shore are always interesting since they are filled with many brightly coloured anemones and the delicate fronds of less common red algae.

Other shores

The commoner types of plants and animals are much the same on other shores, but there may be fewer species. Along the southern part of the east coast of England many of the less common brown algae are absent, while in the north of Scotland, plants seem more obvious than on southern shores, and species of limpets and barnacles are reduced. There are, of course, many local variations, which are partly a matter of substratum or habitat. For example, along the western coasts of Ireland, parts of Scotland, north Devon and north Cornwall, the barnacles of the middle part of the shore are replaced by large numbers of small mussels (*Mytilus edulis*), (Fig. 6) which never reach the economic size of their brethren in the estuaries. This mussel belt is sometimes believed to be due to the proximity of sand, the scouring effect of which may somehow alter the balance between the mussels and the barnacles which compete for space on the rock. A more exact relationship with sand is that of the honeycomb worm, *Sabellaria alveolata*, (Plate V) which builds galleries of tubes from coarse grains of sand and small fragments of shell, cemented onto the rock. Obviously the worm must have a supply of these materials to construct its tube, and it is always found in largest numbers

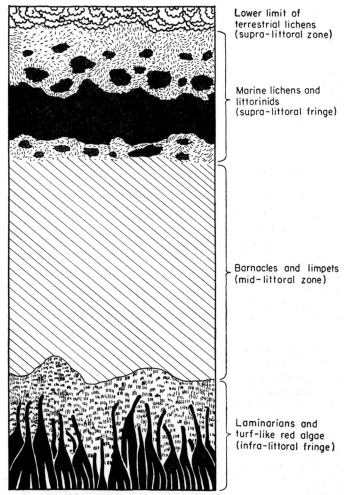

Fig. 17. Very diagrammatic representation of the three zones of the universal rocky-shore system, as seen in a place in S. Devon, on the open coast but sheltered from excessive wave action by outlying reefs.

where rocky reefs emerge from a sandy beach (but only along the western coasts, e.g. north Cornwall and south Wales).

At a similar level on the shore (about MLWN) where sand is absent there may be a belt, or patches, of the white calcareous tubes of serpulid worms (e.g. *Pomatoceros* (Fig. 6) and *Spirorbis*), but this development is rather local, though in other parts of the world it may be a regular feature of the lower shore.

In spite of the great variations that do exist there is a general pattern underlying life on rocky shores, recognisable in most places in the world.

Universal features of a rocky shore

There are three basic zones or belts on all shores, inhabited by similar forms of life, though the actual species present change (Table 1 and Fig. 17).

The uppermost of the three is called the littoral fringe, or sometimes the *Littorina* zone, and is referred to in the Mediterranean as the 'étage supralittoral'. It is always inhabited by a few species of marine snails and by encrusting lichens and simple algae, and is the part of the intertidal zone that is wetted by only a few tides in the year, or by waves and splash only. A few barnacles may be present.

The lowermost of the three zones is called the sublittoral fringe, and is regarded as an extension of the sublittoral zone, not always distinct from it. Here the organisms are nearly always covered by the sea, and exposed for only an hour or two on a few days each month. In temperate and cold climates it is the home of the large brown seaweeds of the laminarian type, including *Laminaria*, *Macrocystis* and *Durvillea*. In warmer climates the smaller fucoids *Cystoseira* and *Sargassum* are more typical of this zone, but many other algae may be found, including calcareous forms and, in some regions, dense beds of marine 'grasses' belonging to the pond weed family of flowering plants. The upper edge of the tropical coral reef belongs to this zone, the predominant organisms being true

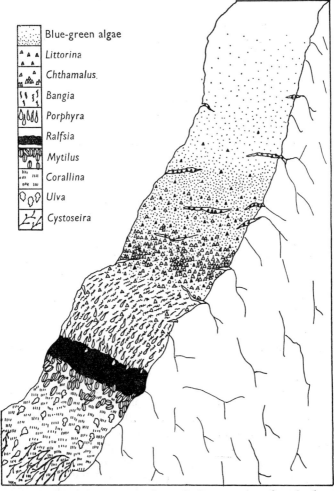

Fig. 18. Pictorial representation of the zonation found along parts of the south coast of France (after Huvé). The vertical height, 1–2 m., is much less than that of preceding diagrams.

stony corals, hydrocorals and calcareous algae, which all contribute to the building of the reef.

The middle part of the shore, called the midlittoral zone, is usually the largest of the three and contains the most characteristic rocky shore organisms. It is the part of the shore that is exposed to the air and covered by the sea every day, and the plants and animals are accustomed to the regular alternation of active and inactive periods. The universal inhabitants of the zone are barnacles, which are absent only in the polar regions where the shore as a whole is barren of life, due to the scouring by ice. The belts of brown fucoids that have been described for Britain are found only in the north temperate regions, and are not characteristic of the midlittoral zone (cf. Fig. 18). In other parts of the world the plant population is very variable, but limpets are almost always present and there is often a subdivision into three smaller belts which are characterised by: barnacles, mussels and calcareous algae with limpets and other barnacles, in that order down the shore.

The characteristic organisms of the universal zones are compared with their British equivalents in Table 4.

Distribution

The foregoing brief discussion of the universal features of life on rocky shores also brings to light some of the differences due to geographical distribution. As in other habitats, many of the plants and animals can live only in a narrow range of temperatures and are thus restricted latitudinally, while others are unable to cross large expanses of ocean and are restricted to certain continents or to one hemisphere. There are however, many other more local discontinuities in distribution of organisms on rocky shores, due to less marked climatic differences, which illustrate similar principles (Figs. 19–21).

In Britain, for example, species of northern or arctic type are most common on, or restricted to, the northern coasts (sometimes north-eastern or north-western coasts). A good example of this type is the dwarf fucoid *Fucus distichus* f. *anceps* (Fig. 20), which is fairly common on wave-beaten shores of the Orkney and Shetland Islands, less common along the outer Hebrides, and has its southern limit in Co. Kerry, Ireland. A larger group of species of tropical or warm temperate distribution is found only on the southern and south-western coasts of the British Isles, and examples are also shown in Figs. 20B, 21A. Analysis of distributions like these suggests the importance of relatively minor climatic factors, such as the occurrence of warm winters or cool summers, on geographical distribution. Studies of this type are easier upon a rocky seashore because the vertical range of the habitat is so small that the distributions are practically linear, and fewer observations are needed than in other habitats. At the same time, care is needed to determine the species correctly, since there may be two or more organisms competing for the same niche and often they are either close relatives or else they show a strong superficial resemblance to one another.

Some rarities of rocky shores

Interesting though it is to note differences and similarities in the zonation and distribution of the commoner organisms, most of us enjoy searching for the rarer forms of life. Apart from the excitement of such searches, the finds generally illustrate some biological principle. As a general rule it is exceptional for a species living on the shore to be classed as a 'living fossil', i.e. a lone representative of a group much commoner in past geological ages and better known as fossils from the rocks, the coelacanth for example. From this we might deduce that competition for survival is keener on the shore than in other habitats. However, a possible exception might be made for the king-crab, *Limulus*, of some American shores, though this is more strictly part of the mobile fauna of sublittoral sands, and perhaps the small brachiopods that have been recorded intertidally from some rocky shores (e.g. New Zealand). The latter are shore representatives of a group that has persisted almost unchanged from Palaeozoic times.

There are many forms that are rare because their habitat requirements are rather strict. The beautiful little solitary

Table 4
Rocky shore zonation

Names of zones

SUPRALITTORAL FRINGE

EULITTORAL OR MIDLITTORAL ZONE	UPPER
	MIDDLE
	LOWER

SUBLITTORAL FRINGE

Typical plants and animals	Typical species found in Britain and approximate tide-levels of boundaries of zones
	EHWS
periwinkles and blue-green algae. Isopods.	*Littorina neritoides* *Littorina saxatilis* *Verrucaria maura* *Ligia oceanica*
	MHWS
periwinkles, barnacles, limpets	*Littorina saxatilis* (*Littorina neritoides*) *Chthamalus stellatus* *Patella vulgata* *Pelvetia canaliculata* *Fucus spiralis*
	E(L)HWN
Barnacles, limpets, small algae including fucoids	*Chthamalus stellatus* *Balanus balanoides* *Patella vulgata, P. depressa* *Fucus vesiculosus* *Ascophyllum nodosum*
	E(H)LWN
Large barnacles, or limpets or tube-worms, with lithothamnia, Corallina, many red algae	*Balanus perforatus* *Patella aspera* *Spirorbis borealis* *Sabellaria alveolata* *Pomatoceros triqueter* *Lithothamnion* *Corallina* *Laurencia* *Fucus serratus* *Himanthalia*
	MLWN to MLWS
Large brown algae (kelps), cystoseiras, marine grasses	*Laminaria* *Sacchorhiza* *Alaria* *Cystoseira* (out of water)
	to below ELWS

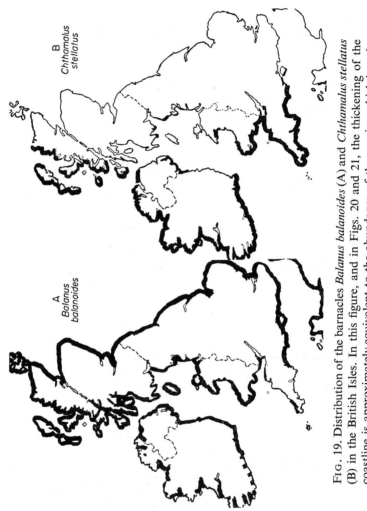

FIG. 19. Distribution of the barnacles *Balanus balanoides* (A) and *Chthamalus stellatus* (B) in the British Isles. In this figure, and in Figs. 20 and 21, the thickening of the coastline is approximately equivalent to the abundance of the species, which is to be regarded as absent in regions where the coastline is thinnest.

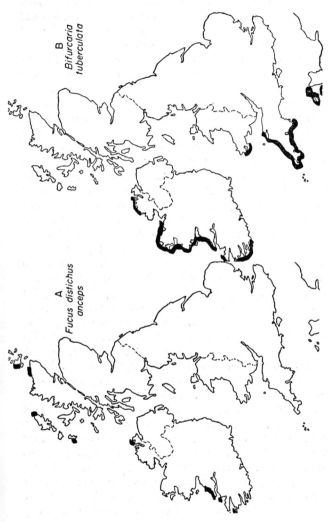

FIG. 20. Distribution of the algae *Fucus distichus anceps* (A) and *Bifurcaria bifurcata* (B). The former is found only on steep wave-beaten rocks facing north, often of difficult access, and may be more continuous than shown here.

corals of our shores (*Caryophyllia smithi* and *Balanophyllia regia*) are good examples of this. Though essentially sublittoral forms, they were once found abundantly in the sublittoral fringe at particular places on the south-west coasts, e.g. on the limestone at Torquay, and Ilfracombe, N. Devon, and on sandstone at Tenby, S. Wales, but now they are very rare thanks to the combined efforts of 'an army of Victorian collectors' (as they were once called) and the increased pollution of these areas due to urbanisation and industrialisation of the coast. However, some of the corals can still be found in abundance under stones on the shores of deep sea-lochs of the west coast of Scotland, where conditions are stable enough and the water remains clear. Stable conditions, particularly the absence of extreme temperature changes, are probably important to the large purple sea-urchin *Paracentrotus lividus*, which can be such a feature of tide-pools on the west coast of Ireland, but which is practically absent from England (two or three records from the south-west in 50 years; Fig. 21).

In addition to habitat factors there are those connected with nutrition, and many of the finely coloured nudibranch molluscs (Fig. 10) are probably scarce (or appear scarce) for this reason. Some of them feed on one or two species of hydroids only, though the sea-lemon, *Archidoris* (Fig. 10), appears to be more catholic in its tastes, and is reasonably common among sponges on overhanging faces. The tectibranch mollusc *Aplysia* feeds on red and green algae, and like herbivores in other habitats may show astonishing fluctuations between states of rarity and abundance.

The question of rarity is connected with an important ecological concept, that there exists a minimum population density below which a species is unable to maintain itself locally without fresh introductions from an outside source (e.g. larvae carried in the water from another shore). This concept is regarded to some extent as psychological in mammals and other higher animals but physiological in the lower invertebrates (e.g. the difficulty of cross-fertilisation in widely spaced sessile animals such as barnacles), but there is evidence

Fig. 21. Distribution of the barnacle *Balanus perforatus* (A) and the sea-urchin *Paracentrotus lividus* (B) on the shore. A few single records of the latter on the English and Scottish coasts are shown.

that all animals are gregarious to some extent and may benefit from crowding.

Mobile animals

Many of the animals of a rocky shore move up and down with the tide, and their numbers can hardly be judged from a survey made at low tide. This category includes edible and shore crabs and most of the shore fishes, but also the wrasses and rocklings which are more consistently sublittoral at low tide and hardly ever allow themselves to be trapped on the shore. It is known that even the sea-urchins move up the shore to feed during high tide (including *Echinus* and *Paracentrotus*), but the extent of any of these movements is little known and offers a field for future work. The larger types of these animals can be observed while wading, with the aid of a water telescope (a long piece of tubing with a watertight glass window at the bottom end and an eyepiece aperture at the top end). In summer, or in warmer climates, snorkel tubes or aqualung gear will offer greater freedom of movement to the observer. It is then possible to investigate the territorial behaviour of fishes, which is akin to that well known in birds.

An opposite movement to that of mobile marine animals is found in those terrestrial animals that go down onto the shore to feed at low tide. Man and rats are omnivorous, while sheep seem to favour *Ascophyllum* and birds concentrate on molluscs. The well-known marine wood-louse, *Ligia*, is best regarded in this category, as an adaptation of a terrestrial form, rather than as a species now evolving in the sea.

Nutrition of life on rocky shores

Many of the animals on a rocky shore feed on the large and small seaweeds that clothe the rocks. The limpets, top-shells, littorinids and sea-urchins are good examples of this type, and all but the urchins feed in exactly the same way, though on different types of plants, by scraping them with a file-like tongue or radula (Fig. 23). The limpets and the top-shells such as *Gibbula* can eat a wide range of plants, from the film of diatoms that often covers the rocks, to quite large fucoids which may be nibbled to the midrib or eaten through at the base (Fig. 23). *Littorina littoralis* favours the large fucoids, while *L. saxatilis* and *Monodonta* are more frequently found on barer rock from which they presumably remove the smaller plants and newly settled young. The sea-urchins also feed by scraping, using five very strong teeth, with which they can tackle anything from encrusting calcareous algae to large laminarians. Strictly speaking they are omnivores and will also feed on encrusting sponges and tunicates, as well as dead fish and other carrion.

The grazing activity of limpets, littorinids and sea-urchins is an important contribution to erosion of the rock surface. Particles of the rock are removed with the food, and the loss in this way has been estimated to be as much as 0·5 to 1·5 mm. a year.

A second type of feeding, filter-feeding, is exemplified by the barnacles and by the small lamellibranch molluscs that live in seaweed and crevices. They strain the water near them and remove any edible particles from it, whether plant or animal or detritus, and then sort it so that only edible matter is ingested. For example, the barnacles sweep the water with their thoracic appendages, or cirri, in such a way that a flow of water is set up over the animal (Fig. 22). Each rhythmic cast of the cirri sweeps part of the flow and any particles large enough (e.g. more than 0.03 mm. in diameter) are caught in fine hairs and are brushed off and collected by smaller cirri. Here they are sorted out and selected fragments are passed to the mandibles and palps to be chewed and swallowed. Sometimes, if the rate of flow of water due to tidal currents or waves is very strong, the barnacles do not beat their cirri rhythmically but simply hold them out in the water flow and withdraw them at intervals to remove captured particles.

The filter-feeders do not greatly influence the growth of the plants of a rocky shore since they feed principally on plankton, and only remove eggs and spores of plants. The limpets and

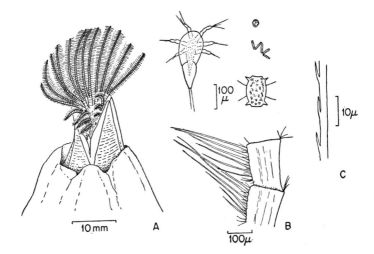

Fig. 22. Filter-feeding method in barnacles: A, view of large barnacle with cirri fully open and just about to be swept forward; B, magnified view of one cirral ramus from the side showing the hairs that strain off food from the water, and some typical food organisms—larva of copepod, large and small diatoms—to same scale; C, very highly magnified view of one hair (seta) to show tiny hooks (setules) that help to hold the food particles.

other rock grazers, however, are very important in the economy of the shore. Experiments show that they control the amount of seaweed growing on the shore, and that where seaweeds are absent it is due largely to the abundance of limpets. This is very easy to demonstrate on practically any shore by removing the limpets from a large marked patch and observing the subsequent appearance and growth of plants. For an experiment of this sort the patch should be at least 5 m. wide, to minimise the effect of movement of surrounding limpets. All the limpets are removed by scraping and smashing with a hammer, paying particular importance to small ones and

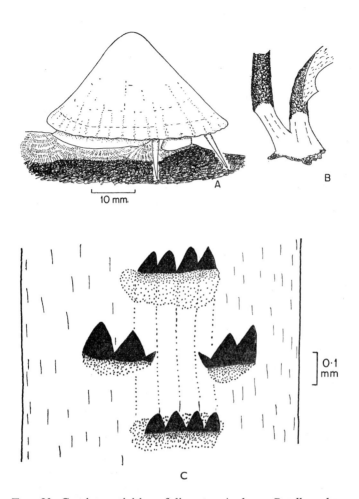

FIG. 23. Grazing activities of limpets: A, large *Patella vulgata* feeding on a coating of filamentous green algae growing on a stone, showing the tooth marks left by the radula in the area already cleared. The mouth is closely pressed to the rock between the two tentacles. B, young plants of *Fucus vesiculosus* showing stipe and mid-rib partly eaten through by a limpet. C, enlarged view of the teeth on radula of a limpet.

those in crevices. The following succession of plant growth may be observed after clearance:
  (i) a few weeks later—a brown film of diatoms;
  (ii) after a few months, depending on the season—complete cover of green algae, chiefly *Enteromorpha*;
  (iii) after one or two years—a dense stand of fucoids.

Eventually there will be a luxuriant growth of fucoids, the species depending on the tide level at which the clearance was made. After two or three years, however, it will be seen that the fucoids are thinning out, and inspection beneath them will reveal a new population of limpets which is preventing new growth and steadily eating away the old. These limpets mostly arrive by settlement of young ones under the thick cover of fucoids. Within five to eight years the original bare rocky shore with limpets and barnacles will have returned.

Experiments of this sort are best begun in the autumn, say September or October, so that the winter and spring are available for the settlement of the plants. Less settlement can take place in the heat and drought of summer, and if a patch of rock is cleared and sterilised (e.g. with a blow torch or flame gun) in March, there is a good chance of heavy settlement of barnacle larvae (*Balanus balanoides*), and the young barnacles will develop to the exclusion of practically all other organisms.

The engraving below shows the rocky shore at Wembury, Devon, and a view of the Mewstone Rock.

4

Sandy and Muddy Shores

In addition to those features of the shore environment already discussed, there are two important factors which influence the animals living in sandy and muddy beaches. One of them is physical and concerns the size of the particles which make up the deposit, the other is the content of organic matter which, with the related sulphur cycle, can be regarded as a chemical factor.

Particle size

The predominant particle size of a beach can be judged quite well by rubbing some of it between the fingers, when it can be classed as coarse sand, fine sand, muddy sand or mud. These categories correspond quite well with the main groups into which terrestrial soils and beach deposits are usually divided during 'quantitative analysis' (see Chapter 8);

Grade	*Diameter of particles (mm.)*
coarse sand	2·0–0·2
fine sand	0·2–0·02
silt	0·02–0·002
clay	less than 0·002

Particles larger than 2 mm. are regarded as gravel, and for some purposes the coarse sand and fine sand grades may be subdivided.

Some workers have preferred what is called the Wentworth scale of grade analysis, in which each grade separated is a regular fraction (half) of the preceding (e.g. 1·0 mm., 0·5 mm. 0·25 mm., 0·125 mm. and so on). Obviously this method separates more grades than a factor of ten and is most useful where the particle composition of the sand varies widely.

Instead of tabulating the results of an analysis, the composition of a deposit may be expressed graphically as a cumulative curve of percentage composition, and the shapes of such curves can be highly characteristic (Fig. 24). For other

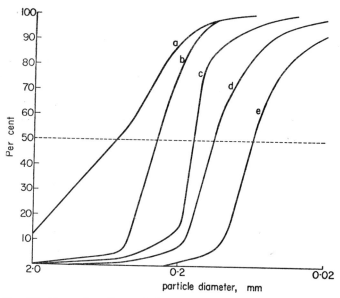

Fig. 24. Examples of cumulative curves, illustrating the results of mechanical analysis. A is a coarse-sand beach, B a fine-sand beach, and C, D and E represent increasing proportions of mud and silt in a muddy sand beach. The median particle diameter can be obtained from this graph by reading along the line for 50% composition.

purposes only the median particle diameter may be quoted, particularly if only small samples have been analysed by direct measurement under the microscope.

The composition of a beach depends mainly on the amount of wave action it receives, but proximity to sources of shingle, sand or silt may also be important. Coarse sand beaches are generally found on open ocean coasts where granitic or similar rocks are exposed and eroded, all the smaller particles being swept away by the waves and tides. On such beaches the water drains away quite quickly at low tide and there may be very little animal life. Beaches of fine sand occur inside bays on the open coast, or along coasts where the prevailing winds do not blow on shore, and generally have a better fauna. Sandy shores grade imperceptibly into muddy sand beaches with increase from shelter from the waves, as for example in sheltered inlets and artificial harbours. Muddy shores depend mainly on complete shelter from the waves but also require a steady supply of fine particles of silt and clay. Most often muddy shores are found in and near the mouths of estuaries, and the fine particles are derived from clays and other suspended matter brought down by the rivers from the land, and precipitated on meeting sea water. Other muddy shores may be found in fully marine conditions inside drowned river valley systems or where there is an outer guard of islands or skerries along the coast.

The particles on a beach are continuously resorted by water movement, particularly by wave action, and this, together with the predominant particle size, determines the angle of slope or 'profile' of the shore. The steepest slopes are found on shingle and coarse sand beaches, and these are the types of shore that are most affected by the destructive forces of the plunging type of wave (p. 13). Such wave action steepens the upper part of the slope and produces what is called the winter profile (Fig. 4). During calm weather or under the action of non-plunging constructive waves the beach profile is smoothed out by re-deposition of material near high-water level. Ripple markings (Plate VII) appear to be formed under the same conditions by altered orbital motion of the water.

Sand dunes form above high-water mark on many beaches. They start around any sort of obstruction on the surface, such as stones, flotsam or plant material, which is sufficient to retain sand blown in shore by the wind, and eventually may build up to 40 feet or more. Further shoreward movement of the sand is usually arrested by growth of marram grass, which may occur naturally or be planted, and which stabilises the deposited sand with its extensive root system and is able to encourage continued dune development. Although dunes cannot resist wave action or strong winds they act as a reserve of material which is available for building up the beach again by favourable wave action or off-shore winds, and without dunes a beach would erode much more rapidly.

Both sandy and muddy beaches retain water when the tide falls, held by capillary action between the particles. The amount of this interstitial water varies according to the size of the particles, but in mixed deposits containing fine sand it is usually about 20 to 25% of the wet weight, falling to 10% or less if there is much gravel or small stones mixed with the sand. On some very soft muddy shores however, the surface layers may contain over 50% water. A rough idea of the water content and particle size of a sand may be obtained by observing whether the material liquefies under immediate pressure of the foot (thixotropy) or whether it appears to solidify, the latter condition indicating considerably lower water content. Obviously, burrowing animals will find life easier in the former type of deposit, which will help their progress through the beach.

Organic content

The particles of a sandy beach are usually quartz fragments derived from the harder rocks, but shell remains are also present, and on most shores there are fragments of organic matter. The organic matter is formed predominantly from decaying seaweeds, but animal remains, faecal matter, and detritus from the land may also contribute. That is, most of the organic matter comes from elsewhere than from the beach itself, and hence more organic particles are found in deposits close to extensive stretches of rocky shore carrying large seaweeds, or where woods come close to the waters edge, and near human habitation. The organic particles are lighter than the quartz and shell fragments and tend to block the interstices of the deposit and thus bind it together.

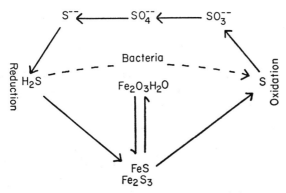

FIG. 25. The sulphur cycle in a sandy beach (after Bruce).

The actual content of organic matter, sometimes expressed as percentage of organic carbon (see Chapter 8), is believed to be one of the main factors governing the density or abundance of the animals living in the deposit, particularly of those species that feed by burrowing through the sand or mud and passing large quantities of it through their intestines (see p. 65). Unfortunately it is difficult to assess the organic matter accurately, and even more difficult to distinguish between utilisable materials which the animals can digest, and non-utilisable materials such as lignin, coal, coke and faecal products. The non-utilisable materials also react with the usual oxidising agents that are employed in the various methods for determining organic content, a fact with which many investigators seem to be unfamiliar.

The organic content of a beach also influences the sulphur cycle in the deposit, by reducing the circulation of interstitial water and by encouraging the growth of bacteria, thus depriving the deeper layers of any oxygen. In all deposits, however, there is a depth at which anaerobic conditions develop, and at which ferric oxide (a normal constituent of beaches) is reduced to sulphide: this point is marked by a change in colour from the surface brown, yellow or grey to a dark grey or black, often smelling of H_2S when exposed to the air (Fig. 25). The position of this so-called 'black layer' is a useful index when comparing conditions on beaches. It may vary from only a few centimetres below the surface in a sheltered place where much organic matter is present, to several decimetres on an open wave-beaten coast. Not much animal life exists below the black layer, though the presence of black deposit at times in the casts of the lugworm *Arenicola* shows that this species can exist and feed under such conditions.

The hydrogen ion concentration of the interstitial water is influenced to some extent by the sulphur cycle and by the organic content, and the most acid values will be found in water withdrawn from the black layer. If this water is allowed to stand for some time in the air it will become opalescent by oxidation of the sulphide to colloidal sulphur. *In situ* oxidation would continue to sulphate (Fig. 25). That the pH falls so little under anaerobic conditions is due to the buffering effect of the carbonate present in the shell fragments. This buffering effect can be shown by bubbling carbon dioxide through a sample of sea water until the pH is lowered. If beach sand is then added and the mixture stirred for several hours in a closed vessel the pH will slowly rise.

The fauna of sandy beaches (Plate VII)

Most if not all of the animals of a sandy beach burrow in the deposit, and to study them thoroughly needs more care than the simple observations that can be made on a rocky shore. Since they have to be dug out and sifted from the sand it is preferable to do so in a way that will allow comparison of the numbers found on different beaches or at different tide levels—that is, a fully quantitative study. However, if time is short, or if no equipment is available, quite a few animals can be captured by scooping out several handfuls of sand and

PLATE III. Some animals from the upper part of a rocky shore: *top*, top-shells, *Monodonta lineata*, in a slight crevice, and scattered barnacles, *Chthamalus stellatus*; *bottom*, at the edge of a small pool and partly shaded by brown seaweeds: sea-anemones, *Actinia equina*; striped top-shells, *Gibbula umbilicalis*; periwinkles, *Littorina saxatilis*, and *Littorina littoralis*; and edible winkles, *Littorina littorea*.

PLATE IV. More high-level animals and plants: *top*, part of a nearly vertical rock face showing the sharp upper limit to the barnacle zone and a distinct narrow belt of channelled-wrack just below it. Scattered tufts of the lichen *Lichina pygmaea* are present among the barnacles further down; *bottom left*, plants of channelled-wrack, *Pelvetia canaliculata*; *bottom right*, barnacles, *Chthamalus stellatus*; limpets, *Patella vulgata*; and lichen, *Lichina pygmaea*.

waiting for the resulting hole to fill with water draining from the surrounding sand. The sand from the bottom of the hole and from the margins is then puddled in the water by continued stirring until some of the animals begin to show themselves. The first will probably be polychaete worms, which will swim up to the surface when disturbed; crustaceans will also swim up and then be trapped in the surface film, while molluscs can be felt with the fingers. An examination of this kind, supplemented

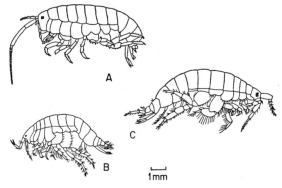

FIG. 26. Some amphipods found in sandy beaches: A, *Talitrus saltator* from HWS; B, *Urothoe* sp. and C, *Bathyporeia* sp. both from MTL to LWS.

by surface observations of the presence of burrows, casts and tubes, may show whether a fuller survey by the methods given in Chapter 8 would be worthwhile.

On some sandy beaches a walk down the slope towards low water will reveal many indications of life below, with three zones of burrows, castings and tubes. Near the extreme high water line, marked by dried cast-up seaweed, we may find little openings in the dried sand, less than 5 mm. in diameter. Some of the holes may be air holes produced by air expelled from the sand the last time it was wetted by the tide, but others lead to the burrows of amphipod crustaceans, notably species of

Talitrus and *Talorchestia*. If there is time, particularly if we are waiting for the tide to fall, a very interesting experiment on visual orientation and habitat selection can be carried out with *Talitrus saltator* (Fig. 26). This is the bigger of the two most commonly found on British shores, and can be captured by scraping down about 5 cm. into the wetter subsurface sand. The amphipods are then taken down the shore to about half tide and placed on the wet sand, the observer standing a metre or more away to avoid interference. After a few moments crawling about, the animals will start leaping up the shore towards high water, covering up to a metre in one leap, until they reach the drier sand above high-water neap level, where they at once start to dig in.

The talitrid amphipods are scavengers that feed mainly on decaying seaweed, for which some of them make excursions down the beach during low tide. Their ability to regain the correct level has been shown to be due to recognition of the skyline and other features in the locality from which they came. It also appears that they can make use of the sun's altitude, the plane of polarisation of light in the sky, and even the moon's position, in association with some internal 'clock', to find their way back from the land if they venture too far up from the tide line. On tropical shores a similar zone is occupied by ghost crabs (*Ocypode*), which return to the water only occasionally to moisten their gill chambers and, of course, for breeding. In some places their burrows extend well above extreme high-water mark (e.g. in the Florida coastal swamps).

The middle region of a sandy beach, from high-water neap to low-water neap, is usually dominated by the casts of the lugworm *Arenicola* (Plate VII), which become larger as we move further down the beach, where they may be mixed with the projecting tubes of maldanid and spionid polychaetes. This region of the shore also carries a dense fauna of amphipods (*Bathyporeia* and *Urothoe*, Fig. 26) and isopods (*Eurydice*), which leave little trace on the surface and can only be found after much stirring.

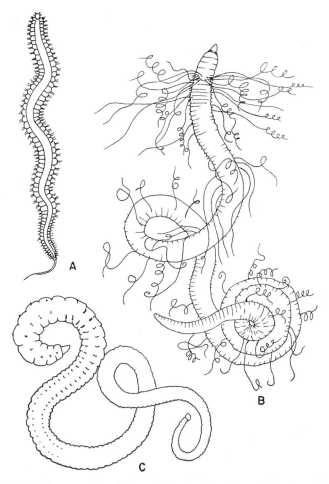

FIG. 27. Some worms from muddy sand and mud: A, *Nephthys hombergi*, MTL down; B, *Cirratulus cirratus*, MTL down; *Capitella capitata*, HWN–MTL down; A is shown about natural size, B and C enlarged about 2 or 3 times.

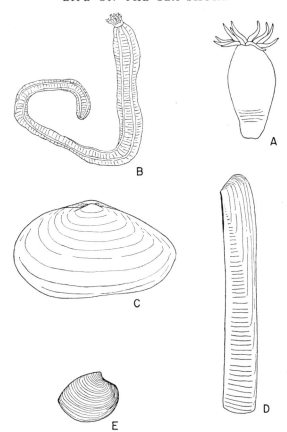

FIG. 28. Some animals from LWS, sandy shores: A, the burrowing anemone, *Peachia hastata*, shown contracted as usually dug out; B, the holothurian *Leptosynapta inhaerens* (from a preserved example); C, large clam, *Mya arenaria*; D, razor-shell, *Ensis siliqua*; E, small clam, *Venus striatula*. (A and B natural size; C, D and E reduced to $\frac{1}{2}$.)

At and below low-water neap level a sandy shore becomes much richer in life, as shown by the variety of castings, burrows and tubes appearing at the surface. In addition to *Arenicola* and other polychaetes, several molluscs and one echinoderm (*Echinocardium*) will probably be found. The most striking mollusc is the razor-shell, *Ensis* (Fig. 28), which leaves an oval-shaped opening to its burrow. These particular holes are easy to find by walking about heavily on the surface of the sand; the animal feels the vibrations and withdraws further into its burrow, throwing up a spurt of water as it retreats. To dig up a razor-shell requires very quick work with the spade, as the more it is disturbed the deeper the animal retreats into the sand, into which it can penetrate very fast. The in-shore fishermen who use these shellfish as bait for long lines, catch them with a short barbed spear which is thrust down the burrow and gives the animals less time to retreat. The American clam shovel is a combination of spade and trowel with a long, narrow blade, and is very effective in skilled hands.

Typical worm tubes of a sandy beach are those of *Lanice*, formed of coarse particles cemented together, but in sheltered places in the south-west the parchment-like tubes of *Chaetopterus* may be found. Other sand-dwelling animals to be dug out at low-tide levels are the worm-like sand cucumbers (*Leptosynapta*, Fig. 28), even more worm-like hemichordates (*Saccoglossus*) which nearly always break up when handled, and the free living anemones (*Peachia* (Fig. 28), *Halcampa*). Other anemones are attached to stones below the surface of the sand (*Cereus*), and the various clams (*Mya*, *Venus*) (Fig. 28) and cockles (*Cardium*) may also be encountered. The smallest molluscs (tellinids) are best developed on fine or muddy sand where they may extend up towards mid-tide.

On tropical sandy shores the lowest zone seems to be dominated by *Emerita* (a burrowing crab), but in temperate regions related species such as *Corystes* are found only sublittorally as adults.

Apart from the predominantly burrowing life, the fauna of a sandy beach includes several more mobile forms. The sand-eels (*Ammodytes*) can often be found just beneath the surface of the sand at low water, while shrimps and small shore crabs

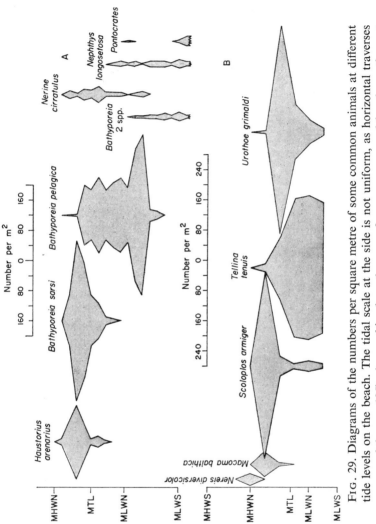

FIG. 29. Diagrams of the numbers per square metre of some common animals at different tide levels on the beach. The tidal scale at the side is not uniform, as horizontal traverses were used (see p. 125). A, amphipods and worms from a coarse-sand beach on the east coast of England; B, a muddy sand shore with estuarine influence. The abundance of *Tellina*

(*Carcinus maenas*) are more widely distributed down the beach, and may not be seen without sifting. Both of the latter have a marbled colour pattern which they can quickly vary and thus merge perfectly with their background.

Although this outline may be found to fit many shores, there may be an extensive variation in numbers and species of animals found in sandy beaches, related in part to varying exposure to wave action, particle size and organic content, and it is impossible to discuss them all here. Some examples of the kind of distribution that may be found on a beach are given in Figs. 29 and 30, as 'kite' diagrams of numbers per unit surface area. The reader is urged to dig his own beach quantitatively and to construct his own diagrams as the best way to understand this problem.

The fauna of muddy shores (Plate VIII)

Beaches that contain mud are obviously more sheltered from wave action than sandy shores, and hence offer a more stable environment. This stability is borne out by the large numbers of free ranging Crustacea and Mollusca that may be found on the surface of many muddy shores, or just within the deposit, or in wet hollows or under stones that may be lying about. Large specimens of *Carcinus* and numbers of the edible winkle (*Littorina littorea*) can often be found, with prawns and shrimps (*Leander*, *Crangon*) in the wetter places. In contrast, of the animals that live within the deposit, a good proportion inhabit permanent burrows and are less mobile than the equivalent forms of sandy shores.

The upper level of a muddy shore, around high-water neaps, may be comparatively barren, unless fresh-water seepage occurs, when the ragworm *Nereis diversicolor* may be common, with small specimens of a spionid worm occurring seasonally. Below high-water neaps the lugworm *Arenicola* (Plate VIII) is common, but is accompanied by several other worms, of which *Capitella* and *Cirratulus* (Fig. 27) may be the most numerous. The active amphipods of sandy shores are replaced by the semi-sedentary *Corophium*, with a permanent burrow.

In the lower half of the zone the tubes of maldanid worms may be very numerous, and digging will disclose species of bivalve molluscs such as *Tellina*, *Macoma*, *Scrobicularia*, *Venerupis* and *Cardium*.

Below low-water neaps the tubes of the large peacock worm, *Sabella*, may be fairly obvious, with beds of the marine 'grass' *Zostera* showing just at the water's edge if the tide is low enough. Sometimes a muddy sand shore becomes predominantly sandy towards low water, when many of the forms already listed above may be found, as well as larger 'sublittoral' stragglers such as the heart-urchin *Spatangus*, the polychaete *Amphitrite* and, in some south-western areas, the scallop *Pecten*.

Some typical population distributions of muddy shores are shown in Figs. 29B and 30B, and again the reader is urged to dig his own shore and construct his own diagrams. As a general principle the fauna becomes more restricted in species and more specialised the more mud is present, and very sticky mud may not repay the trouble of taking samples. It may also be very difficult to move about on, requiring a small boat as a base for operations, or else the use of a mud sledge ('horse') or pattens (like snow shoes), as employed by in-shore fishermen of the Severn and Wash areas when tending their nets and lines.

The flora of sandy and muddy shores

Apart from unicellular algae such as diatoms and dinoflagellates, which are better treated as interstitial forms (p. 69), few plants are found on sandy beaches. Seasonal growths of the brown seaweeds *Chorda filum* and *Laminaria saccharina* may occur where there is some gravel, since they can anchor themselves on quite small pebbles and do not need solid rock for settlement. They are found only at low water, occupying the same level as the laminarians on rocky shores, and are likewise the upper edge of a population extending sublittorally. The green alga *Enteromorpha* may occur attached to small stones further up the beach, but it is better developed on

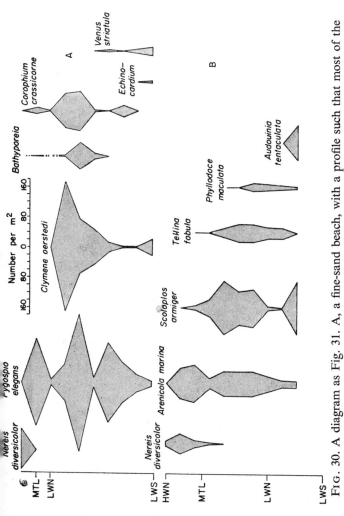

Fig. 30. A diagram as Fig. 31. A, a fine-sand beach, with a profile such that most of the samples on the traverse were taken around LWN. Large numbers of the isopod *Eurydice pulchra* were also found, with a maximum at LWN, but the numbers were too many to be shown on this scale; B, a sheltered muddy sand beach showing a selection of species to illustrate their different vertical distribution.

muddy shores where it may form loose mats lying on the surface of the mud. *Ascophyllum mackai* is an unattached dwarf form of *Ascophyllum* that is found in a similar manner in the very sheltered sea lochs of the Scottish and Norwegian coasts.

On shores where some mud is present the sublittoral fringe is occupied by the flowering plant *Zostera marina*, which can form quite extensive beds or meadows. Other species may be found in other parts of the world, but all appear to have the property of increasing the deposition of silt, which accumulates around the root systems. *Zostera nana*, which occurs up to MTL, but which is less generally distributed, is one of the first colonisers in the development of a salt marsh.

Habits of life on sandy and muddy shores

The animals present in the sand and mud have varied methods of winning their food, and of adjusting their life to the substratum. Like the organisms in other habitats (e.g. terrestrial soil fauna; woodland plants) they live at different depths in the deposit according to their size and mode of feeding. This can be shown by digging out successive layers and sieving them separately, when it will be found that the smallest animals occupy the few upper centimetres of deposit (e.g. the amphipods and capitellid worms) (Fig. 31). The larger worms and the molluscs, which have the ability to reach the surface by active movement or by means of extensible tentacles or siphons, can live much deeper, while only the very largest worms, such as *Arenicola* and *Nephthys*, and the razor-shells and large clams, may be found at more than one spit deep.

Nutrition

The few active carnivores of sandy and muddy shores are species of *Nereis*, *Nephthys*, the phyllodocid polychaetes and the nemertines, but some of these are better regarded as omnivores and scavengers. That is, most of the true predators are external migrants, chiefly the shore fishes, crabs and

SANDY AND MUDDY SHORES 65

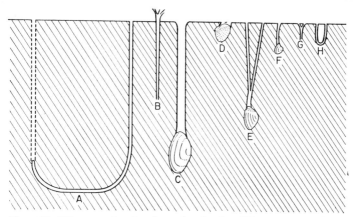

FIG. 31. Diagram showing how deeply the different types of animals burrow in muddy sand and mud; depth shown is about 30 cm. Worms: A, *Arenicola marina*; B, *Lanice conchilega*; G, *Pygospio elegans*; Crustacea: H, *Corophium volutator*; molluscs: C, *Mya arenaria*; D, cockle, *Cardium edule*; E, *Scrobicularia plana*; F, *Macoma balthica*.

wading birds. The remainder of the deposit fauna feeds in three ways, of which the first two are most characteristic. The *deposit feeders* ingest large quantities of the sand or mud while moving through it or maintaining their burrows, and digest any available organic matter. The lugworm *Arenicola* is a good example of a largely non-selective deposit feeder, and when dug up its gut is always filled with sand. It lives in a U-shaped burrow (see Fig. 32) with a well-defined tail shaft and a rather more loosely shaped head shaft which opens into a funnel-shaped depression on the sand surface. A current of water for respiration is drawn in by movements of the worm from the tail shaft and passes forward, while other movements keep the head shaft open, from which the looser material is ingested. From laboratory studies it has become clear that *Arenicola* alternates bursts of feeding and shaft-clearing activity with periods of irrigation, in a regular pattern. This behaviour can be observed by enclosing the worm in a narrow U-tube containing sand or in a very narrow tank made from two sheets of glass clamped together separated by a length of rubber tubing. Smaller tanks of the same sort can be used to follow the behaviour of capitellid worms, which appear to live in a similar manner.

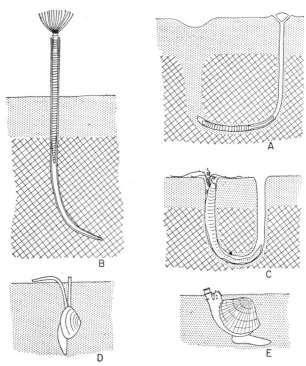

FIG. 32. Feeding habits of some sand- and mud-living animals as seen in sections of their burrows A, *Arenicola marina*, deposit feeder; B, *Sabella pavonina* suspension-feeding worm; C, *Amphitrite johnstoni* selective deposit feeder (surface detritus); D, *Tellina tenuis* another selective deposit feeder; E, *Cardium edule* suspension feeder. The cross-hatching represents the black (sulphide) layer.

Selective deposit feeders have some special collecting mechanism that enables them to remove particles of organic matter (including small animals and diatoms) from the surface of the sand or mud, which is not itself ingested. The terebellid worms and tellinid molluscs are good examples of this type of feeding (Fig. 32C and D). The former spread out a network of ciliated tentacles from the opening of the burrow, over the deposit, and particles picked up are carried along ciliated grooves in the tentacles to the mouth. The tellinids remain well within the deposit, but have extensible siphons (tubular extensions of the mantle openings) which are protruded onto the surface of the sand and mud and moved about, picking up the smaller and lighter particles in the general inhalent flow of water to the gills. The particles are filtered off on the gills as in other bivalves, after further selection inside the shell cavity. The remaining group of *suspension feeders* is not particularly characteristic of depositing shores, as the type of feeding is found in animals from many other habitats. Suspension feeders have adaptations for passing a large volume of water through gills or other devices which filter off small particles (chiefly phytoplankton and the smaller zooplankton), which are then re-sorted internally before being digested. Typical examples of this group are the cockles (Fig. 32E), which appear to use the sand or mud simply as a means to escape predators. Oysters and mussels feed in much the same way without burrowing into the sand or mud.

Respiration in burrowing animals

Most of the burrowing animals are subjected to anaerobic conditions at low tide, either because they close their shells or tubes, or because they cannot extract enough oxygen from the reduced amount present in the interstitial water at any depth in the deposit. Adaptations to deal with this situation are discussed in Chapter 6. All the animals can either irrigate their burrow when they are covered by the tide again, or have mechanisms for passing a current of water over the gills, or have extensible tentacles or branchiae which can be protruded above the surface of the beach.

Mobility of the fauna

Although any tendency of the sand and mud fauna to move up and down the beach with the tide might be considered as a general adaptation to life on the shore, it is best treated here as an adaptation peculiar to burrowing animals. Unfortunately it is a subject in which little work has yet been attempted, although this leaves possibilities for fruitful future research.

The best-known example of tidal movement is the sand crab *Emerita* of tropical and sub-tropical shores, which leaves its burrow and advances up the shore with the rising tide, from low water up to about mid-tide, retreating again as the tide falls. This animal moves with the advance and retreat of breaking waves on the beach, and is able to burrow into the sand and re-establish itself between two wave crests. It is highly likely that a similar movement of the smaller amphipods and isopods takes place on temperate sandy beaches, since these animals show a very patchy, and apparently erratic, distribution when dug out of the sand during low tide. A movement not unlike that of *Emerita* has been found in a Japanese and American species of *Donax*, a tellinid mollusc. This bivalve has been observed moving a short distance up the beach when stimulated by wave wash on the rising tide, but other species of the same genus appear to be more sedentary in habits.

Many of the molluscs are very rapid burrowers; the foot is narrowed and thrust into the deposit, then inflated with body fluids to act as an anchor while the muscles contract to pull in the rest of the body, the process being repeated until the animal is buried. Some worms, such as *Arenicola*, burrow in a similar manner, employing an extensible proboscis in the same way as the molluscan foot, but others, e.g. *Nephthys*, thrust in the narrowed and sharpened head end, with the remainder of the body braced against the thrust. It is surprising how fast many of these animals can disappear if they are dropped onto the sand. Interesting experiments might be carried out, both on the shore and back at the laboratory, to measure the time taken for an animal to dig itself in again, using the same sand or mud at the same temperature for each species.

Even such a large form as the heart-urchin *Echinocardium* is not completely sedentary, and is believed to change its burrow and to travel on the surface of the sand at night. Many other beach animals, particularly the amphipods and isopods, leave their burrows in darkness to swim about in the plankton, and may be captured at night in a net at the water's edge. Here they may be accompanied by multitudes of mysids and other sublittoral sand-living animals, and are preyed on by many species of fish, both young and adult (clupeoids, gadoids, mullet and sea trout, for example).

On British shores it is not easy to follow any tidal movement of the animals living in sandy beaches. On other shores where the tidal range may be less, the observer can remain on the shore and watch the movement take place over a relatively short distance. However, it is quite easy to show that the crustacea leave their burrows and swim about at night at high tide, by sampling the plankton at the water's edge. A simple net for this purpose is D-shaped, towed on the end of a rope by wading or from a small boat, and fished so that the straight edge of the net glides freely over the surface of the sand. The same net can also be used in deeper water or at other times to sample plankton near the surface of the water, by fishing it the other way up, with a float of cork or foam plastic fastened to the straight edge. This type of net and an ordinary simple plankton net are shown in Fig. 44.

The interstitial fauna

During the past twenty years increasing attention has been paid to the very small animals that live in beach sands and gravels. Gradually it has been realised that comparable microfaunas are widely distributed in many habitats; for example, in off-shore deposits, lake and stream gravels, and in many types of subterranean waters. The adaptation common to all these microfaunas is that to life *between* particles. In addition to small physical size many of them are very thin and extremely elongated (Fig. 33), while others are laterally or vertically compressed.

This adaptation to life between the particles illustrates the fundamental difference between the interstitial fauna and the typical macrofauna of a sandy or muddy beach. The latter are 'burrowers' which force their way through the deposit, and thus tend to favour the finer grades, particularly those of thixotropic nature (see p. 148). The interstitial microfauna species are 'sliders' that move about entirely within the pore spaces (interstices) of the deposit. For this reason the large pore spaces of a moderately coarse sand provide the best examples of microfauna, while the more muddy deposits, especially those where the median particle size is below 0·02 mm., are less densely populated. Of course, there is an extensive protozoan fauna, particularly of ciliates, in fine deposits down to a median particle size of 0·01 mm., but in finer grades than this only burrowing forms can exist.

The interstitial animals are believed to feed by grazing on the films of unicellular plants, including diatoms and bacteria, that grow on the particles of sand, but so far very little attention has been paid to this aspect of life in the interstitial habitat. Unicellular plants, including dinoflagellates, can in fact be quite abundant in sandy beaches, at times sufficiently abundant to colour the surface brown.

Nearly all major invertebrate groups have been reported from the interstitial habitat. Representatives of the Turbellaria, Nematoda, Gastrotricha, Archiannelida and harpactid copepods are probably the most numerous, though some of the more extreme cases of adaptation to the habitat are found among modified amphipods and molluscs (Fig. 33). Particle size is the major factor controlling the interstitial fauna, but in addition, the height of the water table in the beach may be important in allowing extension towards high water. This appears to be particularly true of the almost tideless shores of

SANDY AND MUDDY SHORES 71

the Baltic, where some of the animals occur seasonally well above mean sea-level.

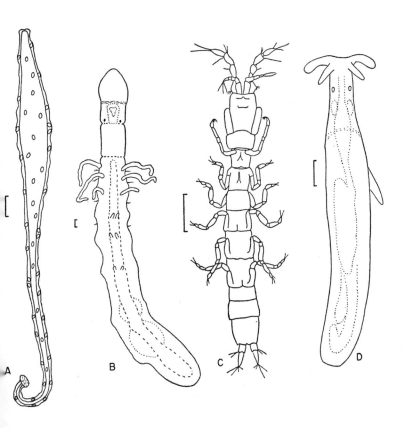

FIG. 33. Examples of inhabitants of the interstitial region, from four different phyla, showing small size and elongate shape. The scale alongside each represents 100μ (0·1 mm.). A, a coelenterate, *Protohydra*; B, annelid, *Psammodrilus*; C, arthropod, *Microcerberus* (amphipod); D, mollusc, *Microhedyle* (opisthobranch gastropod).

Mixed types of shore

Not all shores will fall neatly into the two major groups of eroding and depositing types, and many beaches will be found where there is a mixture of sand or mud or both with outcrops of rock and heaps of boulders. In such places we may find the richest understone fauna of all, with many examples of the typical rocky shore cryptofauna where the mud does not fill all the spaces between the stones.

However, there is an additional group of animals which are adapted to life in pockets of sand or mud between stones. Among the commonest species in this habitat are the two tailless lugworms, *Arenicolides branchialis* and *A. ecaudata*. The former occurs in coarser deposits, from above MTL down to LWN, the latter in the softer and more organic deposits at around LWN and below. Where much organic matter is present and the 'black layer' very close to the surface these lugworms may be replaced by large numbers of the cirratulid worm *Audouinia tentaculata*. All three species occupy temporary burrows beneath stones, the opening lying near to the edge of the stone. In addition to such deposit feeders, several scavengers and carnivores may be found, for example *Sthenelais boa* (Fig. 37) and *Arabella iricolor* (both large polychaetes) as well as the ubiquitous shore crab *Carcinus maenas*.

An extension of this specialised fauna may be found in the small pockets of silt that are present in cracks and crevices on rocky shores. The species in such places are quite varied, with the sabellid worm *Potamilla*, the eunicid worm *Lysidice*, small sipunculids (*Golfingia minutum*) (Fig. 37), and the free-living stage of a semi-parasitic crustacean, *Gnathia*. All of these are easily obtained by levering off cracked layers of rock with a crowbar or heavy screwdriver.

Universal features of sandy shores

Possibly because they have been less generally studied, or perhaps because there is a greater variation in the fauna, it is difficult to establish how universal are the main zones found on sandy shores. According to some authorities, the crustaceans of sandy shores can be divided into three zones, which are more or less equivalent to the three major zones of rocky coasts. There is an upper supralittoral fringe occupied by talitrid amphipods and, in the tropics, ghost crabs, comparable to the periwinkle zone of rocky shores; and there is a lower sublittoral fringe containing a varied fauna with sublittoral affinities, most marked in warm water areas by populations of the burrowing crab *Emerita*, and in temperate latitudes by some amphipods. The midlittoral zone, equivalent to the barnacle zone of rocky shores, is much more difficult to define, since the isopods that are typically present are restricted to a few types of sandy beach. In Europe this is essentially the lugworm zone, but elsewhere the related species are less abundant or else prefer a lower tidal habitat. The midlittoral of muddy sand shores might be called the zone of small clams, since examples of the genera *Mya*, *Macoma* and *Venerupis* are very widely found. In Europe these forms are less obvious in the presence of the more numerous lugworms and other polychaetes.

5

Estuaries and Lagoons

BROADLY speaking, estuaries and lagoons are an extension of the depositing shore habitat, with an added complication that the salinity is lower (estuaries) or occasionally higher (lagoons, some peculiar estuaries—see below) than the sea. Solid substrates are rare and may be confined to a few rock outcrops, piers and pilings, whereas very extensive expanses of soft mud may be exposed at low tide.

An estuary is usually a river valley which has been invaded by the sea, owing to a rise in sea-level or a sinking of the land. Typically it has the shape of an elongated triangle, with sea water penetrating from the wide end and fresh water flowing in from the narrow end. The action of tides against the flow of the river produces a very complex mixture of salinities which varies according to the shape of the estuary and the rate of flow of fresh water and tide. Except in some aberrant tropical estuaries there is usually a gradient in salinity, increasing downstream towards the sea. There is often an accompanying vertical gradient, with the lighter fresh water flowing out at the surface and the heavier sea water flowing in beneath (see Fig. 34). However, in some estuaries where great turbulence occurs there may be only the horizontal salinity gradient. A further variation in salinity, attributable to the rotation of the earth (Coriolis force), is that higher salinities may be found on the left-hand side (looking towards the sea) of estuaries in the northern hemisphere, and on the right-hand side of estuaries in the southern hemisphere; a good example of this is found in Chesapeake Bay, a very large estuarine complex on the east coast of America.

An organism living in an estuary may experience daily or twice daily changes of salinity of considerable magnitude.

For example at a point 12 miles up the Tamar from the sea, where the truly marine species reach their limits, the salinity can vary from 25‰ at high tide to as little as 4‰ at low tide, occasionally dropping to 1‰ at low tide after rainy weather. In addition to salinity changes there is usually a greater range of temperature, and the water is nearly always charged with large amounts of silt and clay in suspension. It is not surprising

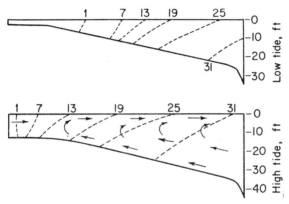

FIG. 34. Very diagrammatic longtitudinal sections of an estuary at low tide and at high tide to show salinity distribution (as parts per 1000). Depth is given in feet below the water-line. The arrows show the suggested pattern of circulation, with incoming sea water flowing along the bottom and gradually mixing vertically with the outgoing surface stream of fresh water.

therefore that the number of species successfully adapted to life in the upper reaches of estuaries is small, though these few species may be individually very numerous.

Investigations into the penetration of organisms from the sea are quite easy to make, though a small boat will often be found useful in reaching the edges of mud flats and other less accessible spots. It is usually best to choose a number of fixed places or stations (e.g. those easily reached by car) at intervals down the estuary, and travel in an upstream direction so that the animals being observed first become less common, then rare, before vanishing completely. At the same stations it is usual to take samples of the water at high tide and at low tide, during dry seasons and rainy seasons. The salinity of these samples, which can usually be determined quite simply by the hydrometer method (see Chapter 8) will give an idea of the daily and seasonal changes experienced, and allow some sort of map of salinity distribution to be prepared. Ideally, a boat should be used to take samples in midstream and towards the sides, and from the surface and the bottom water (see Chapter 8), so that sections of salinity distribution can be drawn (see Fig. 34).

Where it has been found that there is a vertical salinity gradient in an estuary, with more saline water at the bottom, we might reasonably expect marine forms to penetrate farther upstream on the bottom in the centre. But, since there is tidal fluctuation as well, it is more often found that the high-level shore species are in fact the ones that penetrate farther from the sea, since they are exposed only to the more saline water present at high tide. On the other hand, the mud does tend to retain the higher salinity water in its interstices, and thus burrowing animals may be at a lesser disadvantage than more mobile or rock-living species.

As a rule very few algae penetrate the upper reaches of an estuary, and the laminarians and all but a few red species drop out within one or two miles of the mouth (Fig. 35). The fucoids are more hardy and may in fact reach lower salinities than many marine animals that can protect themselves with a shell (e.g. barnacles). In the Baltic Sea where salinity is low, but does not fluctuate, both *Fucus vesiculosus* and *Balanus improvisus* are found at salinities as low as 4 or 5‰. A few genera of green algae may do very well indeed in estuaries, and the mud banks may be clothed with dense mats of *Enteromorpha*, *Chaetomorpha* and *Cladophora*. Sometimes these species are accompanied by a resistant red alga, *Gracilaria*. The salt-marsh plants are described later.

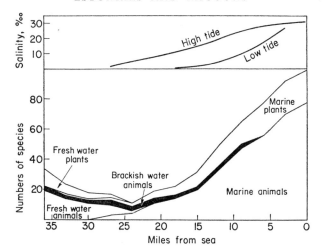

FIG. 35. Penetration of marine organisms up an estuary; number of species plotted against distance from the sea. The graph above shows the distribution of average salinity at high and low tides (River Tees, after Alexander, Southgate and Bassindale).

The majority of estuarine animals are derived from the sea and very few (chiefly insect larvae, oligochaete worms, and pulmonate gastropod molluscs) are of fresh-water origin. A third group of species comprises those evolved from marine forms but which have become specially adapted to life at reduced salinities. A classical form of diagram illustrating these three groups is given in Fig. 36. The brackish water forms are the true estuarine animals, and are the most interesting, many of them being found all over the world, either as single species or groups of species belonging to single brackish water genera. To mention a few, there is the hydroid *Cordylophora*, the anemone *Diadumene*, the errant polychaete *Nereis diversicolor* (and other species), the serpulid polychaete *Mercierella* (Fig. 37), the well-known amphipod *Gammarus* (several species), and several mysids and prawns. Among them are many interesting examples of species which have a fresh water or a marine 'twin', with little overlap in the salinity distribution of the pairs, e.g. the sub-species of *Gammarus zaddachi*; *Mysis oculata* (marine) and *Mysis relicta* (brackish, but see p. 85); and species of the prawns *Palaemonetes*. In certain cases, such as *Gammarus duebeni*, a species is capable

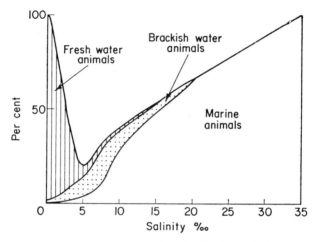

FIG. 36. Diagram illustrating the distribution of fresh water, marine- and brackish-water species as a function of average salinity (N. Europe and Baltic area). Given as percentage of the species penetrating from fresh water and percentage of marine- and brackish-water species penetrating from the sea.

of living in both brackish and fresh waters in the absence of a competing fresh-water species (*G. pulex*).

The brackish-water species perhaps illustrate one of the presumed pathways of evolution of fresh-water animals from the sea, through a fairly tolerant brackish-water form which later became separated into two distinct species by further adaptation of a fresh-water form.

Lagoons

Marine lagoons, enclosed bodies of water with salinities higher than the sea, are found only in warm climates with low rainfall. They are cut off from the sea by sand bars or chains of islands and are usually regarded as being formed by inundation of the land. Typical examples are found in parts of the Black Sea and Sea of Azov, along the Gulf of Mexico and in South Africa, and the bitter lakes of the Suez Canal system are similar in character. The salinity may indeed be very high, even exceeding 100‰, and few animals and plants can exist. The successful organisms are those of estuarine or brackish-water type, since true euryhaline species can tolerate both low and high salinities (see p. 83). In addition, some lagoons also have elements from the fauna and flora of salt pans, e.g. the brine shrimp *Artemia salina*, which are apparently of freshwater origin.

It is believed that many of the animals find the salinity of the lagoons too high for successful breeding, and that the

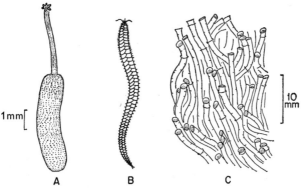

FIG. 37. A, the sipunculid, *Golfingia minutum*, from the special crevice fauna; B, the worm, *Sthenelais boa*, from pockets of gravel under stones; C, the tube-worm, *Mercierella enigmatica*, found in brackish water in the south, especially where warmed by power stations, etc.

populations are maintained by larvae brought in from outside by currents. Hence those lagoons that are permanently cut off from the sea have a very reduced fauna, restricted to a few Crustacea and hardy fishes.

Hypersaline estuaries

In some aberrant estuaries, usually found in arid desert regions, the rainfall may be so low or seasonal that more water is lost by evaporation than is gained from rainfall. Hence the middle reaches of such estuaries may have salinities higher than that of the sea outside. Some colder estuaries with very small fresh-water drainage may also show a seasonal hypersalinity of less extreme nature (e.g. River Crouch in Essex).

Salt marshes

The development of a salt marsh appears to depend not on low salinity but upon shelter from wave action and the presence of a continued supply of silt and clay particles. However, these factors are rarely found on the open coast, and salt marshes are best considered here, as part of the estuarine habitat.

The earliest colonisers of a mud or muddy sand flat are the blue-green algae, the green alga *Enteromorpha* and, less commonly, the flowering plant *Zostera nana*. These forms constitute the seaward fringe of a developing marsh, from low water up to MTL. At this stage the deposition of mud is fairly steady, with a few obvious drainage channels in which there may be an extensive marine fauna, including prawns, shrimps and young fish such as flounders, dabs and plaice, for which estuaries sometimes provide a nursery ground. Above this primary zone the true salt-marsh flora is present, with extensive beds of *Salicornia* up to HWN and beyond. On some coasts, as on the south and west coasts of England, the earlier stages of the salt-marsh development may be speeded up by growth of *Spartina townsendi*, which is a very rapid coloniser, and causes heavy accumulation of silt around its extensive root system. Near HWN *Salicornia* tends to give way to *Glyceria* and *Aster tripolium*, and the accumulation of silt becomes thicker. *Suaeda maritima* may also be present, as well as free-living masses of the brown alga *Pelvetia canaliculata* and of the red alga *Bostrychia*. At normal HWS level a much greater variety of plants begins to appear, with the marsh much cut up into patches by a multitude of drainage channels. Sea-lavender (*Limnonium*) may be common on some coasts, with *Spergularia*, marsh arrow grass (*Triglochin*), thrift (*Armeria maritima*) and *Obione* (=*Atriplex*). The highest levels of the marsh are flooded by the sea on only a few occasions, and silt deposition ceases; here plantains (*Plantago*) and *Artemisia* may be found. The higher levels of a salt marsh have an essentially terrestrial fauna, with insects, mammals and birds predominating.

Mangrove swamps

The tropical shore equivalent of a salt marsh is the mangrove swamp, which can likewise develop on a fully saline coast but is more usual in sheltered bays, inlets and estuaries where the salinity may be less. There is again a regular zonation from the sea to the land, reflecting a succession of colonising plants. The primary colonisers of the bare sand or mud flat are the red mangroves, *Rhizophora*, which are very well adapted to this task. Their seeds germinate on the parent plant and drop into the mud ready to take root, and in this way avoid being smothered or drowned in the vulnerable embryonic stage. As the marsh builds up towards high water the red mangroves begin to die off, apparently because of lack of oxygen in the deposit. The black mangroves (*Avicenna*), which take over, have a specialised root system with aerial pneumatophores (projections of the roots containing air cavities) above the surface of the completely anaerobic mud. Above high-tide line button-woods replace the mangroves, and ultimately a tropical swamp forest is formed (e.g. the Florida Everglades). The intertidal roots and stems of the mangroves and the pockets of organic mud held between them provide a special habitat for many tropical animals, which may be rich in both species and numbers. Typically there is an abundance of rock oysters and barnacles attached to the roots and stems of the red mangroves, with many mussels and more mobile animals among them (e.g. hermit crabs, *Clibanarius*; fiddler crabs, *Uca*; several gastropods and the mud-skipper fish, *Periopthalmus*). The latter and one or two of the gastropods and other animals, are regarded as having travelled part of the way along the line of evolution to a terrestrial life. It is thought that the mangrove swamp equivalent in Carboniferous times (various ferns and clubmosses) may have played an important part in the colonisation of the land from the sea.

Adaptations to estuarine life

Variation in salinity appears to be the greatest hazard to life in estuaries, and many species survive by resisting the change. Barnacles and burrowing molluscs can close their shells during the low salinity period, and open up again later; experiments suggest that *Mytilus*, for example, will close at 25‰. Soft-bodied forms are not so fortunate, and some have to take refuge at the upper tidal levels, where they are wetted only by the more saline waters. In each case, however, the adaptation is similar to that needed to withstand life out of water, and comes naturally to an animal already fitted for life on the shore. These *resisters*, as we may call them, are marine forms in which the salt content of the blood or the body fluid is kept close to that of the sea.

The most successful estuarine animals are those that have learned to tolerate or accommodate themselves to the wide range of salinities of the external medium. The *tolerant* species are generally found among the lower groups of animals. Their body fluids have a salinity which follows that of the estuary without apparently causing any great derangement of their metabolism, any adjustment taking place at the cell level. The higher forms of estuarine life, including the active species, show *regulation* at epidermal or cuticular level, and are able to adjust their body-fluid salt levels independently of the external medium, by close control of the uptake and excretion of water and salts. The degree of such salinity regulation varies from species to species, and its estimation is one of the classic experiments on estuarine life.

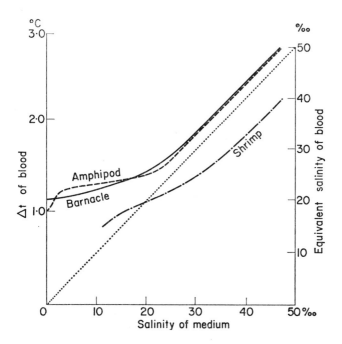

FIG. 38. Graph showing the salt content of the body fluids of three species of crustacea, as equivalent salinity and as depression of the freezing-point (Δt.) in relation to the salinity of the external medium. The amphipod (*Gammarus duebeni*) and the barnacle (A form of *Balanus balanoides* from the Murman Sea) show the regulation typical of brackish-water species; they are at all times hypertonic (blood salt concentration greater than sea water). The shrimp, *Crangon crangon*, is a coastal species that also lives in estuaries, but is not as well adapted as the other species.

For example, we may take a sample of shore crabs (*Carcinus maenas*), and another of edible crabs (*Cancer*) or swimming crabs (*Portunus*), and divide each sample into three lots placing one lot in normal sea water, another in sea water diluted with an equal volume of glass-distilled water, and the remainder in some sea water concentrated by evaporation, or by the addition of some dry sea salt ('Tidmann's'), to bring the salinity up to about 150% of normal. After some hours, or longer, a sample of body fluid is removed from each crab, by cutting a hole in the shell, and its freezing point determined. From the freezing-point depression we can calculate the percentage sea water in the body fluid and compare it with that of the medium.[1] The edible crabs and the swimming crabs may have a body fluid concentration close to that of the medium, and may in fact die before the experiment ends. The *Carcinus*, which can tolerate life in estuaries well, will be found to maintain higher salt concentrations in their body fluids than in the diluted sea water, and lower concentrations than in the concentrated sea water, and this difference can be maintained indefinitely. That is, the animal strives to maintain a salt level in its blood and tissues near to that of the open sea to which it has been adapted in the course of evolution (see Fig. 38).

Fresh-water species living in estuaries are generally restricted to areas with salinities below 5‰ and a special adaptation is required to fit them for life at higher salinities. In such cases, for example during the breeding migrations of the fresh-water eel and the feeding migration of salmon, the body fluids are kept at the relatively low salt content typical of fresh-water fish, salt is actively excreted from the gills and there may be changes in skin structure to reduce water loss. Marine birds, which drink sea water when away from land, have a mechanism for salt excretion from the tear glands. Further details of osmoregulation will be found in a companion volume in this series.[2]

[1] For smaller species it is necessary to use microscopic freezing-point methods or conductivity measurements to determine the salt concentration in the small amounts of blood available.

[2] Lockwood, P. M. J., *Animal Body Fluids and their Regulation*, 1963.

Reference was made earlier to the brackish-water species *Mysis relicta*. The name of this species connects it with the interesting problem of the so-called glacial relicts, which has some bearing on adaptation to life in estuaries. These relict forms are single species belonging to otherwise essentially marine genera, and are found in certain fresh-water lakes and brackish inland seas over N. Europe, Siberia and N. America. At one time it was suggested that they had evolved from the nearest marine species in an era during the glaciations, when the lakes first communicated with an essentially arctic sea and later gradually became fresh. If this were so, many of the apparently identical populations of Europe and America would have to have evolved independently. It is now believed, on the basis of further work on the lakes, and further collecting in the arctic, that nearly all the 'glacial relicts' are really brackish-water forms, present in parts of the White Sea and along the Siberian and N. American coasts as well as in the lakes. The relict lakes are now thought to have communicated with a large body of brackish water during the last glacial period, when the ice was retreating and many of the European rivers become dammed up against the ice front. Widespread flooding and ingress of brackish water is thought to have occurred, which assisted the spread of the 'relict' forms. These were already adapted to life in brackish waters and little further change was called for as the isolated lakes became fresh. The species survive because of the absence of competition from other animals and because the temperature conditions in the lakes provide a body of cool bottom water during the summer. In addition to *Mysis relicta* the relict fauna includes an isopod, *Mesidotea entomon;* an amphipod, *Pontoporeia affinis;* and a species of bullhead fish, *Myxocephalus* (or *Cottus*) *quadricornis*. These or other members of the group have been found in the Baltic Sea, parts of the Caspian Sea, the Aral Sea, Lake Baikal, and in some of the Siberian rivers as well as in the isolated lakes already mentioned and along the Arctic Ocean coasts.

6

Adaptations to Life on the Shore

As already noted there is much greater daily variation of environmental factors on the shore than in other habitats, and the animals and plants are profoundly modified or adapted to cope with these changes. The most important need is to prevent or reduce the effects of loss of water when exposed to the air.

Resistance to desiccation

Unless specially protected, the tissues of a marine plant or animal will begin to dry out when exposed to the air. Protection from or resistance to water loss is found in many organisms, but quite a few appear also to be able to tolerate considerable loss of water. Among the fucoids, for example, there is a considerable resistance to water loss during short periods. Over longer periods the species from high-tide levels may loose up to 60% of their total water content, but most of this comes from the cell walls, which are very much thicker than in low tide forms, and the cell contents remain undamaged. Some of the simpler red and green algae found near high water (*Porphyra* and *Enteromorpha*) (Fig. 5) can sometimes be found dried out to an almost paper-like consistency on the surface of the clumps, but are only killed in summer when the drying is accompanied by prolonged high temperatures and strong sunlight.

In contrast to the plants, most of the animals found on the upper part of the shore resist drying inside an impervious shell, which can be tightly closed up (barnacles and small lamellibranchs) sealed off by a horny membrane (littorinids) or kept closely pressed to the rock (limpets). In many of these forms there is a direct relationship between shell thickness and

PLATE V. Plants and animals from below mid-tide level: *top*, a mass of tubes of the honeycomb worm, *Sabellaria alveolata*; *bottom left*, serrated-wrack, *Fucus serratus*; *bottom right*, thongweed, *Himanthalia elongata*, growing on rock covered with coralline algae, *Corallina officinalis* and *Lithothamnia*.

PLATE VI. Plants exposed at low spring tides: *top*, mixed *Laminaria digitata* (smooth fronds), and *Laminaria saccharina* (wrinkled fronds), with a few tufts of *Cystoseira baccata*; *bottom left*, a small bed of *Zostera* on a muddy sand shore with stones and gravel at the surface: a few *Laminaria digitata* and *Laminaria hyperborea* (stipes erect) in the background; *bottom right*, a wave-beaten shore, with *Laminaria digitata* and *Alaria esculenta*.

ADAPTATIONS TO LIFE ON THE SHORE

period of exposure to the air, and some very thick shells have been found in *Patella vulgata* from high water, up to 1 cm. thick at the apex of the cone.

Experiments on the resistance of the high-water plants and animals to drying are quite easy to carry out. Dilutions of sulphuric acid of measured density are made up from tables found in the *Handbook of Physics and Chemistry* (see Further Reading).[1]

If these solutions are kept in closed desiccator vessels the air above will have a constant relative humidity. The organisms under test are placed in a series of different relative humidities (e.g. 95%, 75%, 50% and so on) kept at the same temperature, and are removed at intervals in closed vessels for weighing. In this way the water loss can be plotted graphically up to the point where the organism appears dead and fails to recover on return to sea water.

The more mobile animals that cannot adopt protective measures are forced to seek shelter and shade when the tide falls, as for example many shore fishes, worms and crabs, which retreat into tide pools, into deep crevices or down the shore towards low tide. At the middle levels of the shore we may find a few soft-bodied sessile forms that are apparently quite resistant to drying, such as the ordinary dark red and brown forms of the anemone *Actinia*, and the naked hydroid *Clava squamata* on the fronds of *Ascophyllum* in sheltered places. These forms produce a copious secretion of mucus which must in some way help them to resist drying. At the lowest levels of the intertidal zone the animals and plants are exposed to the air for only a short time, and no special adaptation seems to be needed, though even at this level the softer-bodied animals are best developed in crevices and under overhanging surfaces.

The humidity of the atmosphere is obviously important. It is fairly high on the shore at most times, and even in warm,

[1] Some delicate animals are killed by exposure over sulphuric acid solutions; in such cases other compounds can be used, though they are less convenient to prepare.

dry, spring and early summer weather values of from 45 to 60% relative humidity have been recorded close to the rocks, rising to 85 to 93% on cool, moist, autumn days with weak or hazy sun. Under the latter conditions, and during rain or drizzle, exposure to air will cause little or no loss of water and, in fact, both shelled animals and crevice species remain active, feeding and moving about as if they were still covered by the tide. The humidity is also high at night (no measurements are available, however) and the same species may then remain active in comparatively dry weather.

Heat and cold

The thick shell that protects high-level intertidal animals from desiccation offers some insulation against excessive heat or cold. However, neither a thick shell nor thick cell walls can prevent the organism being heated up by sunlight. This heating effect has been measured by means of fine wire thermocouples placed inside the shells of different species (see Table 5 and Chapter 8). On a dull day, when the tide falls the tissue temperatures will rise or fall from near the temperature of the sea until they are nearly in equilibrium with the air temperature, modified by any evaporative cooling that is taking place if the humidity is low and there is some wind. In sunlight however, the tissue temperature begins to rise as soon as the organism is exposed to the air, and will continue until the input of heat is balanced by loss due to conduction and radiation from the body surface. Such heating by direct radiation can be quite marked even in winter, when it can mitigate or completely nullify the effects of low air temperature. In summer radiative heating is immensely more effective, though the temperate latitudes cannot equal the rise found on tropical shores. However, on tropical shores it would appear that some of the animals can cool themselves by allowing a little water to escape and evaporate from the body or shell opening. Generally, water trapped under or inside the shell is used for this purpose, but sometimes water is lost from the tissues. Water lost in this way by evaporative cooling must be balanced

Table 5

Temperatures of animals on the sea-shore (°C)

	Sunny day in summer	Dull day in summer	Sunny day during frost	Dull day in winter
Sea temperature in-shore	14·6	14·3	6·3	7·6
Temperature of tide pools	17·4	14·5	4·9	4·9
Shade air temperature on land	16·5	14·5	0	4·5
Shade air temperature close to rocks	17·8	15·0	3·9	5·8
Temperature of barnacles	26·8	14·7	10·5	6·3
Temperature of limpets	24·8	14·7	7·5	5·4
Temperature of limpets in shade	20·9	—	4·0	—
% relative humidity	60	95	60	80

against the need to avoid desiccation, and there is obviously a limit to the degree of temperature regulation possible.

It appears that cold weather is less severe on the shore than on the land. Firstly, there is proximity to the relative warmth of the sea, and the twice daily immersion in it; secondly, the heating effects of sunlight even for short periods: and shade air temperature close to the rocks during low tide is nearly always higher than shade air temperature measured by a standard screen thermometer a few hundred yards away. Local effects of sunlight and of proximity to the sea were well shown during the cold winter of 1962–3. The cold-intolerant southern barnacle, *Balanus perforatus*, perished at MTL even in the extreme south-west where it was shaded from the sun or in sheltered creeks and harbours, but survived where it was close to the open sea or exposed to sunlight.

Most of the worst effects of the same cold winter on seashore life were found in places where there had been a very excessive fall in sea temperature (e.g. around the Isle of Wight) and the freezing weather was accompanied by easterly winds blowing directly on shore.

Respiratory exchange

As far as can be seen the plants of the shore exhibit no adaptation for gaseous exchange, and probably none is needed since the general diffusion rate should be adequate at most times. Abundant supplies of bicarbonate are available in sea water, and although lack of carbon dioxide may reduce photosynthesis when the plants are out of the water, liberation of oxygen should be possible at all times.

In the animal world, respiratory exchange is difficult and many different modifications and adaptations can be found. As in most aquatic animals, the shore forms take up oxygen and liberate CO_2 through the general body surface or more usually in thin-walled extensions of it (gills, branchiae or tentacles). Such structures are not very suitable for gaseous exchange in the air since they require a continuous flow of water over them to maintain efficiency, and are among the

ADAPTATIONS TO LIFE ON THE SHORE 91

first parts of the body to dry out unless specially protected. Thus there is a tendency for species from high-tide levels to have their gills enclosed in a compartment in the shell cavity, and as a result respiratory exchange is reduced during low-tide periods. The same is true of species living in sand and mud, whether or not they have a shell or merely inhabit a temporary burrow. In all these forms the animal appears to remain quiescent at low tide, and any oxygen debt due to anaerobic activity is quickly removed by vigorous ventilation of the gill cavity, shell or burrow when the tide returns. The tube worm *Sabella* and the errant polychaete *Nephthys* are examples of this type, as are most lamellibranchs. *Nephthys* indeed appears to be cut off from practically all supplies of oxygen at low water, and does not attempt to maintain any aerobic respiration, the haemoglobin in its blood being adapted for oxygen

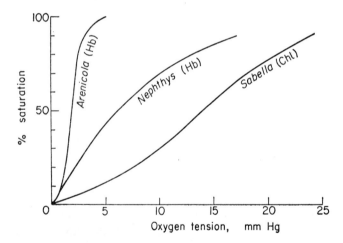

FIG. 39. Oxygen dissociation curves of the respiratory pigment in the blood of three species of polychaet worms. As percentage saturation with oxygen at different external concentrations (tensions). Only *Arenicola* picks up large amounts of oxygen at low tensions and becomes saturated. Hb haemoglobin; Chl chlorocruorin.

transport at high tensions only. The blood pigment in *Sabella*, green chlorocruorin, shows even less facility for oxygen transport in other than fully aerated conditions (Fig. 39).

Other burrowing forms such as *Arenicola marina* show more specialisation. The burrow remains open at low tide and partly filled with sea water containing reduced amounts of oxygen. The haemoglobin dissociation curve (Fig. 39) shows that the animal can usefully transport some oxygen from the burrow to the tissues even at extremely low oxygen tensions. Moreover, if the water in the burrow should become completely stagnant it appears that the worm may move up the tail shaft and draw down again with a bubble of air, thus renewing the oxygen content of the water in the burrow. The old theory, that the haemoglobin in the blood of *Arenicola* is a reserve of oxygen for the low-tide periods is now quite discredited; calculations show that it could not last more than an hour or two, whereas many of the worms are at levels where they will have to withstand 6 to 10 hours with the burrow uncovered by water.

The less specialised forms generally show a linkage between oxygen pressure and oxygen consumption, as in *Actinia*, *Nereis* and *Asterias*, for example, and their activity is reduced under low oxygen concentrations. Indeed, *Actinia* has been found to close up at levels below 2 cc. oxygen per litre.

On rocky shores the species living towards high water show a tendency towards aerial respiration, and some have developed accessory respiratory organs for the purpose. The simplest condition is found in barnacles, where the mantle cavity contains a bubble of air separated from the outside air only by a thin film of water across the partly open entrance to the cavity. At times the shell may be closed, at others it may open wider, sufficiently to renew the air bubble but most of the time it is very slightly open. Most barnacles on the shore will be found to be slightly open if they are approached carefully, but will quickly close up if a shadow falls across the shell or if footsteps cause vibration of the rock. On a quiet day it is possible to hear a clicking noise as one walks across a rocky shore and the barnacles close their shells abruptly.

ADAPTATIONS TO LIFE ON THE SHORE

Limpets usually hold a good deal of water and some air between the shell and the body, and the shell is often held slightly off the rock, allowing exchange of gases to take place. In the littorinids, however, which are out of the water for long periods, part of the mantle cavity is vascularised, and thus more lung-like, while the number of gill or branchial folds is correspondingly reduced. A similar trend is found in tropical hermit crabs and ghost crabs, with a series showing gradual development of a vascularised lung chamber in addition to the gill chamber, and a pumping mechanism for exchange of air. The typical high-tide forms can range some distance inland and may be able to survive complete excision of the gills indefinitely. The mud-skipper, *Periopthalmus*, of mangrove swamps, has an elongated branchial chamber which appears to function as a gas-storage organ.

Feeding

Obviously most of the shore animals can feed only when covered by the tide, particularly the filter-feeding types which strain off small particles from the sea water (e.g. lamellibranch molluscs, barnacles) or the detritus feeders which pick up and sort organic particles from the surface of the sand or mud. The exceptions are the burrowing non-selective deposit feeders, and the predators and rock-grazing forms, though the latter two types can move about on the surface at low water only when the humidity is high enough to prevent desiccation.

The principle adaptation of the filter-feeding group is a rapid response to wetting, whereby they resume feeding immediately they are covered by the tide, so that none of the short time available is wasted. Most of the barnacles, and also some of the lamellibranchs found in weed and crevices at high water, will feed when covered only by a thin film of water, as for example between the splashes of successive waves. This response can be demonstrated on the shore or in the laboratory by simply squirting water over the animals from a bulb pipette or syringe. Many will open in less than a minute, and one barnacle species, *Chthamalus stellatus*, will readily put out its cirral net and show feeding movements (see p. 45).

In this way the filter feeders can extend the apparent period of submergence and can exist at levels that are not covered by all tides. In such forms the metabolism is adapted to alternations of feeding with temporary starvation, and we might expect a daily rhythm related to the tides. In other forms that occasionally feed even at low tide the daily rhythms will be less obvious and may be dominated by a day and night rhythm, such as is found in many intertidal crustaceans (see p. 109).

Excretion

It seems probable that most intertidal forms with a daily or tidal rhythm of their metabolism need no further adaptation to assist the elimination of their waste products, since this will normally occur while they are active under water. However, some of the high-level gastropod molluscs, notably *Littorina neritoides*, may be dry for days at a time, and may feed in damp weather without being covered by the tide. In these forms it has been found that the nitrogen metabolism has altered in a direction also found in terrestrial forms. That is, the proportion of uric acid to ammonium salts is increased, as the following analyses of the contents of the excretory glands demonstrate:

	mg. uric acid/g.
Fresh-water species of gastropods	0·1
Intertidal and sublittoral species	0·5 to 5·0
Littorina neritoides	25
Terrestrial species (*Helix* and *Limax*)	31 to 700

Some high-level barnacles (e.g. *Chthamalus*) continue excretory activity when kept moist out of water and are able to push faecal pellets and their own cast skins (chitinous exoskeletons cast by a form of moult or ecdysis every few days) clear of the shell cavity.

Modifications to breeding and larval development

It is common, but by no means universal, for marine animals to have eggs and sperm which are shed freely into the water, where the egg develops into a drifting, swimming, larval stage that has to feed and grow before it can assume the adult form. Therefore modifications to this pattern are not necessarily to be regarded as adaptations for life on the shore. Many sublittoral species, particularly those of arctic seas, may have some form of internal fertilisation and incubate the young to

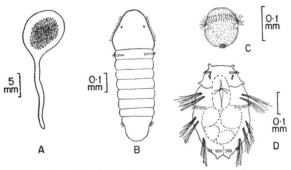

FIG. 40. Breeding of polychaet worms: A, egg capsule of *Scoloplos*, anchored in the sand; B, late larva of *Scoloplos* dissected out from the capsule—no pelagic stage; C, early trochophore larva of *Phyllodoce maculata*, hatched out of egg capsule into plankton; D, later planktonic stage of *Nereis diversicolor*.

a late stage, or else large yolky eggs with direct development to the adult form. Nevertheless, most of these modifications to the life history *are* found among intertidal species, in addition to the forms in which a more extended larval phase appears to be no handicap.

Types of larvae

Thus, although fertilisation is external, many of the larger worms have big yolky eggs that develop directly into a bottom-living stage. Those of *Arenicola* usually lie freely on or in the sand of the beach; the eggs of *Arenicolides*, however, tend to remain inside a portion of the gelatinous lining to the burrow and develop there. This process is carried further in other polychaetes, for example, *Scoloplos armiger* (Fig. 40), in which the eggs are laid in a specially shaped gelatinous cocoon anchored by a stalk in the surface of the sand. In this cocoon the young *Scoloplos* develop directly from the egg into a young bottom-living stage (Fig. 40B). Similar cocoons are laid by phyllodocid worms for example, *Phyllodoce maculata* (Fig. 40C), but from these there hatches a free-swimming planktonic stage which later develops into the adult form. External fertilisation is less common in the molluscs, and the advanced gastropods lay eggs in gelatinous masses (e.g. *Littorina littoralis*) or in horny capsules (*Nucella*) where they develop into small bottom-living stages; more rarely, the eggs are incubated in a specially developed brood pouch inside the parent shell. At one time it was thought that the species of *Littorina* showed a clear correlation of larval type with the tide-level at which the adult lived on the shore, starting with planktonic eggs and larvae in *Littorina littorea* (LWN), through egg-laying and direct development in *L. littoralis* (MTL), to incubation in a brood pouch and viviparity in *L. saxatilis* (HWN). However, researches only thirty years ago showed that *L. neritoides*, the highest of them all (HWS and above) had eggs that were shed into the sea, and that there was a larval phase in the plankton.

Echinoderms are not very common on the shore except at low water, and are generally thought of as having a long larval stage in the plankton. This is true of most sea-urchins and starfishes, which have beautiful and very distinctive larvae (Fig. 42A). However, one common small starfish of northern shores, *Leptasterias mulleri*, not easily distinguished from the common *Asterias rubens*, incubates its young. This form can be found quite high up the shore in some Scottish localities and it is obviously better adapted for life on the shore than other starfish.

Most of the hydroids living in the intertidal zone generally have the normal alternation of generations modified, so that

ADAPTATIONS TO LIFE ON THE SHORE 97

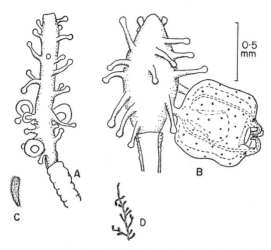

Fig. 41. Breeding of hydroids: A, single hydranth of *Coryne muscoides*, showing gonothecae (spore sacs producing eggs and sperm); B, single hydranth of *Sarsia* (*Syncoryne*) *tubulosa*, with well-developed medusa ready to break off and live in the plankton; C, more highly magnified, planula larva produced from eggs of A; D, natural size of A.

eggs and sperm are produced by the hydroid stage in a modified medusa-bud without an intervening free-living planktonic stage. For example *Clava* (Fig. 10) and the species of *Coryne* may be compared with the sublittoral species of *Sarsia* (or *Syncoryne*) which have a full medusoid stage (Fig. 41). Likewise, in thecate hydroids, the commonest shore species is probably *Campanularia flexuosa*, with direct development, which may be compared with the related *Obelia* species found in the sublittoral, all with a full medusa stage. In both direct and medusoid development the fertilised egg develops into a ciliated planula larva which swims and crawls about for a few hours or days before choosing a place to settle.

However, in spite of the obvious adaptations mentioned above, some of the commonest shore animals have planktonic larvae. Thus, limpets have a well-marked spawning period, some species showing mass spawning of all mature animals within a few days; the eggs develop in the sea into a ciliated larva (veliger), which forms a tiny shell and settles down on the rocks after a free-swimming life of from two to ten days. The common mussel, *Mytilus edulis* (Fig. 42), and most other lamellibranchs, have planktonic larvae which may live for

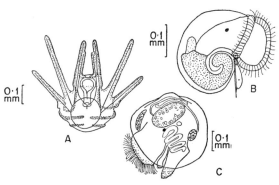

FIG. 42. Larvae of shore animals, taken from the plankton: A, the sea-urchin, *Echinus esculentus*; B, the edible winkle, *Littorina littorea*; C, the mussel, *Mytilus edulis*.

weeks in the sea, though the oyster retains its eggs inside its shell and does not liberate them until they hatch out into ciliated larvae which are free to fend for themselves.

This method of liberating motile larvae from eggs incubated by the adult is more common in crustaceans, and for example, the crabs, lobsters and prawns carry the developing eggs attached to some of the small appendages, or swimmerets, under the abdomen; they are then said to be 'in berry', and cannot be sold for food in Britain but are supposed to be put back into the sea. Some of the lower crustacea may carry the eggs in a brood pouch, and some (e.g. amphipods and isopods) retain them there to final development without any planktonic stage. In barnacles fertilisation is internal and the

ADAPTATIONS TO LIFE ON THE SHORE 99

eggs are laid inside the shell cavity. Here they may be held for weeks or even months before being hatched and set free as a planktonic nauplius stage (Fig. 43). There are six of these nauplius stages, the second within a few hours of liberation, but the remainder only after the larva has had time to feed and grow. The final (7th) larval stage is the cypris, looking rather

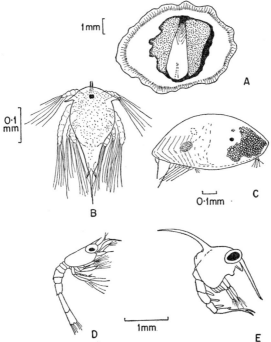

FIG. 43. Breeding of some crustacea: A, underside of a barnacle after removal from the rock, showing the basal membrane partly torn away to reveal the two egg masses on either side; B, second stage nauplius larva of the barnacle *Balanus balanoides*; C, cypris larva of a barnacle about to settle on the rocks; D, zoea larva of the shrimp, *Crangon crangon*; E, zoea larva of the shore crab *Carcinus maenas*.

like a small bivalve mollusc, but in fact named after one of the more common genera of a rather little-known group of crustacea, the ostracods, which have a bivalve shell in the adult stage. The cypris larva sometimes swims and sometimes crawls on the bottom, and eventually selects a place to settle, where it cements itself down and metamorphoses into a tiny barnacle. In prawns and crabs the larvae liberated from the eggs carried by the adult pass through a number of equivalent planktonic stages called zoeas (Fig. 43), ending with a partly bottom living megalopa stage which is more like the adult.

The common shore fishes, e.g. gobies, blennies and suckerfish, lay their eggs in masses cemented to the undersides of stones and in crevices, where the parent remains on guard and helps to keep them healthy by wafting a current of water over them. The eggs hatch out into tiny fish which live in the plankton for some time. These young fish and many of the larvae already described feed at first on small plants of the plankton, but later eat one another and larvae of other groups as they grow bigger and need more concentrated protein food.

Collection of larvae

A great many larval forms can be collected on a visit to the shore, especially in the spring when a high proportion of the worms and crustaceans begin breeding. The larvae can be kept in culture and form the basis of many interesting observations and experiments.

It is probably easiest to obtain barnacle larvae; *Balanus balanoides* in March and April; *Chthamalus stellatus* between June and September. The adults are levered of the rock with a knife, when the pair of egg masses lying underneath should be visible (Fig. 43A). Ripe embryos look darker in the mass because of the presence of a black or red eyespot that develops in each larva at a fairly late stage. If well-developed egg masses are placed in dishes of clean sea water the embryos will begin to struggle and will break through the egg case within a few minutes or hours. They then swim up towards the light and can easily be sucked up in a pipette and transferred to a fresh dish of sea water. After an hour or so the first nauplius moults into the second nauplius stage, but further stages will not emerge unless the larvae are able to feed on cultures of microscopic algae. Some of the lengthy processes in development of larvae, of other animals as well as barnacles, can be bypassed by catching the later stages in the sea, in the plankton. A boat is not needed, and in estuaries and bays many larvae can be caught by streaming a net in the tidal currents from a jetty or pier, or by swishing a net among seaweed as the tide rises (see Fig. 44). If barnacles are breeding, a net used in this way may catch some cypris stages which can be watched under the microscope for their reactions to different surfaces (e.g. stones and shells) and to water currents (e.g. puffs of water from a bulb pipette. Other larvae that might be easy to obtain are those of *Phyllodoce* and related genera of polychaetes with gelatinous egg capsules attached to weed or sand. The egg capsules of *Nucella* will be found under stones, nearly all the year round, and those of *Littorina littoralis* on *Fucus*,

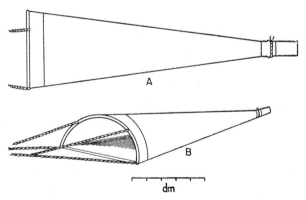

FIG. 44. Plankton nets: A, a simple net for towing from a small boat. The mesh is of nylon bolting cloth, with terylene or canvas sailcloth where it is stitched onto the ring and where the bucket is tied on. B, a similar net fitted to a D-shaped 'ring' for use on the bottom or at the water's edge.

nearly as often. These two will not produce planktonic larvae of course. However, soft ribbons of eggs laid by different nudibranchs at different times of the year will provide ample free-swimming mollusc larvae for experiment. Most interesting of all are perhaps the 'tadpole' larvae that are often freely produced by the ascidians *Dendrodoa*, *Botryllus* and *Morchellium*, or can be teased out from the adults by dissection. These larvae demonstrate the chordate nature of these otherwise undistinguished animals, and possess eyespots, tail and notochord (Fig. 11D).

To obtain early larvae of *Patella*, *Mytilus* and *Echinus*, it is necessary to take ripe eggs from a mature female and fertilise them with active sperm from a male. The shell or body must be opened and a small part of the gonad teased out under the microscope; mature females have eggs or oocytes that readily break free of the surrounding tissue and round off in the water; mature males have masses of active sperm lying free in the canals and ducts. As a general rule male gonads are white, cream or pinkish, and show white smears of sperm when teased out. Female gonads are more often coloured orange, green or brown, and look granular like fish roe. Sometimes a more or less natural spawning will take place in the laboratory, so that only clean, ripe eggs are extruded from the oviduct; this process can sometimes be hastened in *Echinus* by cutting the shell in half and leaving the top half upside down over a jar of sea water. Some times the presence of eggs or sperm in the water will cause some of the other animals to spawn. At other times it may be necessary to break up the gonad and swish it about in water to free the gametes from connective tissue. However obtained, the eggs should be added to jars of clean sea water and a little sperm suspension added (some people add a dense concentration of sperm to a mass of eggs and then wash off and suspend in clean water afterwards). For all experiments of this sort the instruments must be of stainless steel, and the vessels of glass, cleaned in concentrated sulphuric acid and washed in many changes of distilled water. The sea water should be from off shore, and preferably filtered through a

Berkefeld filter candle, which will remove all other larvae, bacteria and other fine particles. The developing eggs should be kept at an even temperature (that of the sea at the time of the main breeding season), the dishes covered to exclude dust, and away from sunlight. Food in the form of algal cultures will probably be needed, though a few species have been reared on fine suspensions of powdered ox liver (a bacteriological reagent), and the later stages of crustacean larvae will eat eggs of *Echinus* and other invertebrates. The echinoderms, however, require plant food throughout, and are almost impossible to rear except under special embryological laboratory conditions.

Larval dispersal

There is obviously no general principle running through the breeding methods of shore animals. Clearly, loss of young stages by too wide a dispersal in the sea during a long larval phase has to be balanced against other factors, and the need to colonise or recolonise distant areas, which can only be done by planktonic larvae, may have a survival value for the species. As a rule, an increase in the size of the eggs or reduction of the planktonic phase should result in a greater survival of the young; correspondingly, the loss that accompanies a longer larval life is usually compensated for by a greater output of larvae. On the whole it would seem that the best adapted intertidal forms have either a fairly short planktonic life or else the eggs are incubated for a while before passing into the planktonic phase. The latter development results in some control over the actual hatching and liberation of the larvae, so that, as in barnacles, the young can be cast free only when there is sufficient food (e.g. diatoms and flagellates) in the water to ensure survival. Some species, such as the tunicates, have a very short larval life during which the larva does not feed, but is purely a means of dispersing the species.

An interesting example of dispersal by planktonic larvae and of change in distribution is found in the spread of the barnacle *Elminius modestus* in Europe (Fig. 45). This species has its home in harbours and estuaries in Australia and New

Zealand. At the end of the 1939–45 war it turned up on the south coast of England and researches since then indicate that it was introduced at an earlier date by ships carrying the adult barnacles on their hulls. Most probably the concentration of foreign trade ships in the early months of the war, while the convoy systems were still being instituted, led to a high population of adults and larvae, sufficient to allow a breeding population to develop around Southampton. From here the species appears to have spread by means of larvae carried in coastal currents, helped in one or two places by further shipping infections, until the animal was present all along the south-east coast up to the Wash and along the European coast almost to Denmark. Second and third sets of infections by ships occurred in the Irish Sea more recently, and the path of these has been traced successfully. In many places there was an advancing 'front' of adults, whose year by year expansion could be related to water currents, of the order of 1 mile per day during the larval period. Where the barnacle did not spread quickly along a coast it was always found that other evidence indicated that water currents flowed in the wrong direction.

This establishment and spread of a foreign species has been helped by its ability to compete well with the native forms. It can tolerate higher temperatures than many native species, and though less tolerant of cold than one of the commonest of them, is able to make up for this by its prolonged and heavy breeding period. It is also resistant to estuarine conditions, and has replaced native species both on the open coast in the south-east (*Balanus balanoides*) and more widely in estuaries (*Balanus improvisus*).

Selection of habitat

There is a most important adaptation that probably occurs in all main groups of animals living on the shore: that of selection of a suitable place in which to metamorphose. Some mobile forms seem always to settle first in pools or even below low water, and only later assume the adult zonation pattern

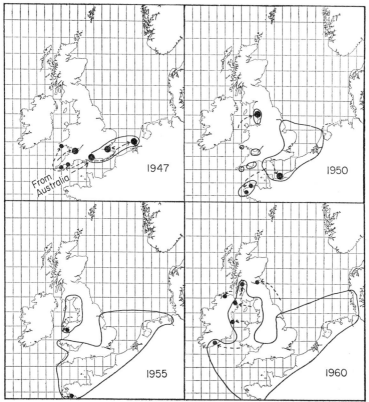

FIG. 45. The spread of the barnacle, *Elminius modestus*, in N.W. Europe. For each map, the main boundary of the species is shown up to that date, together with centres where it was known or believed to have been introduced up to the same date. The original centre of introduction, Southampton Water, is shown as a large black circle, as are five other major centres where later introductions took place. Less-important centres of dispersal are shown as small black circles. Arrows indicate the probable origin of each introduction. Between 1950 and 1955 the species spread to a small area in N.W. Spain (not shown on the map) but there have been no other introductions outside the area of the map. (After Crisp.)

(e.g. *Littorina littorea*, *Patella*), but sessile species must exercise care in metamorphosis, since they cannot move again if by any chance they have chosen an unsuitable place. In these species the free swimming larva seems to be provided with a sense that enables it to choose a suitable site. In the serpulid worm *Spirorbis* some chemical factor from the habitat is recognised, for example something from the fronds of *Fucus* on which the adult lives, or other algae. In other serpulids and in most barnacles an allied phenomenon known as gregariousness is found. This means that the larvae are attracted to and settle in places where adults of the species are already common. Barnacle larvae, for example, are attracted to pieces of rock on which

Table 6

Gregarious settlement of barnacles (after Knight-Jones). In each experiment 20 cyprid larvae of *Balanus balanoides* were placed in a dish of 250 ml. sea water containing one stone, bare or with barnacles as listed.

Substratum provided	No. of larvae settling in 24 hr. in total of 12 experiments
Bare stones	nil
Stones with *Chthamalus*	nil
Stones with *Verruca*	2
Stones with *Elminius*	15
Stones with *Balanus crenatus*	13
Stones with *Balanus balanoides*	71
Stones with empty shells of *Balanus balanoides*	40

In other experiments larvae settled on stones from which even the shells had been removed, but which carried chemically detectable traces of the barnacle shell base.

the adults have been living, but from which they have been removed (Table 6). The response persists even after some chemical cleansing, and has been tracked down to the chitinous parts of the shell. Not only are the larvae able to recognise a place where a barnacle is or had been living, but they can in preference choose a place where the same species, or a closely related species, is present. Such a specialisation of the chemoreceptive sense implies a greater degree of sophistication than expected in otherwise lowly animals, possibly comparable with the senses that guide migrating salmon to the home rivers and streams in which they were spawned.

The breeding of the algae is perhaps more complicated than that of the animals since many groups and species show alternation of generations. The red algae, which have by far the most complicated life histories, might be regarded as being least well adapted to intertidal life, a position to which they would be assigned also on their pigments (see p. 116). Many have three phases: a haploid gametophyte gives rise to a diploid carposporophyte, which in turn produces a diploid tetrasporophyte (e.g. *Polysiphonia*, *Delesseria*). Others have the tetrasporophyte absent or reduced to microscopic size, and both gametophyte and carposporophyte are haploid (*Nemalion*, *Asparagopsis*). Alteration to these types has produced variations in which the sexual and sporebearing phases are combined in one plant which may be haploid, diploid or triploid. Some of the commoner shore species are represented by a tetrasporophyte only (*Lomentaria*), in others the tetrasporophyte phase appears unknown (*Gigartina*). Furthermore, in all red algae the male gametes are non-motile, and fertilisation is dependent on water movements.

The larger green algae show a tendency towards reduction of phases in the life history. Thus, in *Ulva*, *Enteromorpha* and some species of *Cladophora* there is an alternation of otherwise similar plants, one producing gametes, the other spores. In some other species only the haploid gametophyte develops into a plant, and the fertilised egg undergoes immediate reduction division and forms a number of zoospores, each of which gives rise to a new gametophyte. In a few of the larger green algae, such as *Codium*, *Valonia* and some species of *Cladophora*, there appears to be more adaptation to life on the shore, the plant being diploid and reduction division taking place during gametogenesis, the fertilised egg giving rise to the young plant directly.

A clear division of life histories can be found in the brown seaweeds. All the orders other than Fucales have alternation of generations, either between equally sized gametophytes and sporophytes, as in most of the smaller forms, or between a large sporophyte and small, often microscopic, gametophyte as in the laminarians. All these forms are found towards low water or in the immediate sublittoral. The fucoids, which range all over the shore up to HWS, have no alternation of generations. Reduction division immediately precedes gamete formation as in animals, and the egg develops directly into the young diploid plant. However, even the high-level fucoids are dependent on the sea for fertilisation, since the motile antheridia must swim.

Algal life histories are studied in much the same way as the development of some of the shore animals, that is, by setting up cultures of spores or fertilised eggs obtained from a mature plant. The spores or eggs are washed in filtered sea water and allowed to settle onto microscope slides, which are then immersed in covered dishes of sea water enriched with a culture medium (see Chapter 8). It is usually impossible to obtain bacteria-free cultures, but this does not normally matter unless details of the plant's nutrition are being sought. In this case it is necessary to try to rear the plant in a bacteria-free medium to ensure that only the substances present in the culture medium are being used by the plant for growth. The culture vessels should be kept under uniform conditions, preferably in an illuminated, temperature-controlled water bath or culture room. The exact conditions, particularly temperature, degree of illumination, and composition of the culture medium may have to be changed to obtain the best growth to suit each species.

In this way it has been found possible to rear young plants of most of the fucoids, and of hybrids between the species; the gametophyte stages of other brown algae; and the various stages of the red and green species with more complex life histories. Results of researches on these lines allow association of sporophyte and gametophyte phases of the same plant even when of very different form and previously described as separate species, and help to place some plants in their correct systematic positions.

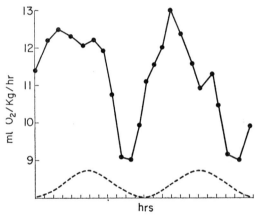

FIG. 46. Rate of oxygen consumption of the fiddler crab, *Uca*, compared with the tidal curve during the period of observations.

Rhythms

Many of the adaptations already described illustrate how the tides govern the lives of the shore organisms, and how their physiological functions and processes show a definite rhythm of activity with a maximum at high tide. This is what is called an exogenous rhythm. However, in many animals it has been shown that the rhythm will persist when the animal is kept continuously under water (Figs. 46 and 48). Similarly, an animal displaying 24-hour rhythms apparently related to the alternation of day and night (Fig. 47) may continue to show the same variation when subjected to continuous light or darkness.

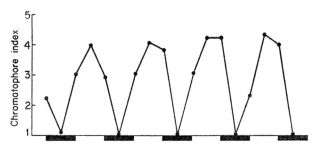

FIG. 47. Diurnal rhythm of expansion and contraction of the black chromatophores (melanophores) in the skin of the crab, *Uca*. Periods of outside darkness are shown in black.

These examples indicate that many organisms must possess internal, endogenous, rhythms governed by some sort of 'clock' in their metabolic system, which can be, and normally is, synchronised with external variables such as light and tide. Whether or not the rhythms are basically internal or are externally controlled does not really affect us here, in considering the widespread nature of their occurrence in so many different types of shore animals. Both tidal and 24-hour rhythms have been found in nearly all major groups (Fig. 48): the turbellarian *Convoluta* comes to the surface of the sand at low water and buries itself again when covered by the tide; *Actinia*, several polychaete worms, oysters and littorinids show higher respiratory rates at high tide. However most investigations have been carried out on Crustacea, to which group many of the findings of workers on insects are applicable. Particularly noticeable and easily assessed changes take place in the black and brown pigment cells, generally in a 24-hour rhythm (Fig. 46). The general term for these cells is chromatophore, and some animals have dark cells (dark red, brown,

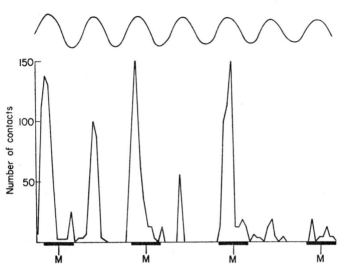

FIG. 48. Tidal and diurnal behaviour rhythms in the crab, *Carcinus maenas*, when kept continuously damp out of water, exposed to dim red light. The vertical scale shows relative activity, as number of contacts with a swinging partition in the tank. Periods of outside darkness are shown in black, M is midnight. The upper graph gives the tidal curve for the duration of the experiment. Most activity was found when outside darkness coincided with high tide.
(After Naylor.)

black) referred to as melanophores, and light-coloured chromatophores with white or yellow pigment. The dark cells are contracted in the daytime, and the light cells expanded, usually in proportion to the amount of light and the colour and pattern of the background on which the animal is placed—i.e. it assumes a tint and pattern that helps to merge its outline into the background. At night the melanophores expand and the white chromatophores are contracted so that the animals become generally blacker or bluish. The changes are quite easy to follow by examining the integument of a crab or prawn under the microscope. An arbitrary index of chromatophore sizes is selected (Fig. 49—'melanophore index') and the animals are allotted to one of these stages according to the state of expansion of the cells. From observations on a large group, twenty or more, at intervals of two hours or so, it is possible to draw up a graph showing the mean daily changes in the average chromatophore condition, as in Fig. 47. Ancillary observations have shown that the chromatophore changes are a general manifestation of some metabolic change and are accompanied by simultaneous daily rhythms in general activity and rate of respiration (Figs. 46 and 48). In crabs and prawns it has been shown that all these changes are under the control of glands situated in the eyestalks near to the eyes. These are ductless glands, and secrete several hormones which are carried in the blood to the various organs and to the chromatophores. Removal of the eyestalks, and hence of the ductless glands, will produce striking changes in the colour pattern of many species, which can be changed again by injection of extracts of the glands. It appears that the rhythms are effected by some reaction of the glands to light reaching the eyes.

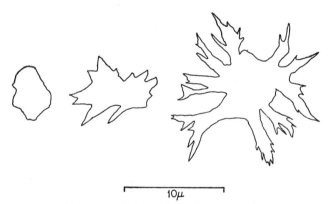

FIG. 49. Highly magnified outline drawings of pigment cells (chromatophores) of a prawn, *Leander serratus*; contracted state on left, two stages of expansion on the right.

ADAPTATIONS TO LIFE ON THE SHORE

In other animals with well-marked colour responses, e.g. squids and cuttlefish, there may be direct nervous control of the chromatophores, and hence much more rapid changes in colour of the animal. The colour changes in an octopus are almost instantaneous, from pale to rich red-brown and back again. Both octopods and cuttlefish can assume various colour patterns which have been shown by psychological experiments to be associated with various moods of the animal and its state of tension or excitement. In cuttlefish, which are not normally found on the shore, but may be taken as juvenile stages in a net close to the beach at night, there is well-marked 24-hour rhythm of behaviour, the animals remaining buried in sand on the bottom in daylight and becoming active and feeding mainly at night.

7
The Causes of Zonation

IT is obvious enough that the zones of plants and animals found on the shore are related in some way to the tides or other changes in water level, and that the various species differ in their tolerance of exposure to air or submersion in the sea. But the changes in water level generate many other secondary factors which are believed to sharpen the pattern of zonation. Thus, the falling tide exposes an organism not only to drying (water loss) which is modified by atmospheric humidity, but to increased radiation (sunlight, including ultra violet) and increased temperatures in summer or low temperatures in winter; supplies of nutrient salts are cut off, and opportunities for gaseous exchange (respiration, photosynthesis) may be reduced. Immersion in the sea mitigates all these effects but produces further problems for a shore organism, including a reduction in illumination and a change in spectral composition of the light as well as an influx of mobile predators from the sublittoral zone.

There has been little attempted investigation of the causes of zonation by experimental methods, and most studies have been made by recording the pattern of zonation in detail and then attempting to relate the zones to tidal factors. For example, much attention has been paid to points on the tidal curve where there is a marked change in the proportion of time spent under water or exposed to air, and where many of the organisms may be found to have upper or lower limits (see Fig. 14). However, some laboratory work has been carried out on one or two aspects of exposure to air or immersion in the sea, by subjecting a species to one of the effects (e.g. heat, drying, reduced illumination, increased ultra violet illumination) that can easily be varied, and then comparing its tolerance with the

THE CAUSES OF ZONATION

variation of the same factor at the tidal level at which it lives. Experiments of this sort have been made on fucoid algae and on molluscs of the limpet, periwinkle or top-shell type. These are all easy to keep in the laboratory, and the fucoids have an additional advantage in showing very sharp sub-zones on the shore. Animals of sandy and muddy shores seem to have been ignored in this way, except for work on the behaviour of talitrids, already described (p. 56).

Immersion and light

Among the common fucoid algae the rate of growth appears to vary with the proportion of time the plant is immersed. In some simple experiments carried out fifty years ago on young plants ('sporelings'), *Fucus spiralis*, which occurs near high water, was found to grow best under conditions of 6 hours wet and 6 hours dry, and grew little or not at all when kept out of water for 11 out of 12 hours or immersed for 11 out of 12 hours. In contrast, *Fucus serratus*, which occurs near low-water neaps, grew best under almost continuous immersion, did not grow when 6 hours immersion was alternated with 6 hours dry, and died when kept out of water for 11 out of 12 hours. The mid-tide species *F. vesiculosus* and *Ascophyllum*, showed some growth both when immersed for 11 out of 12 hours and under equal periods of immersion and drying, but failed to grow or died when kept dry for 11 out of 12 hours. These experiments would be quite simple to repeat on young plants of the same and other species, kept in simple culture, in sea water or enriched sea water, under normal daylight, growth being assessed as change in weight or length.

Many experiments have shown that there is a graded resistance to drying and to ultra-violet radiation, plants from high-tide levels surviving longer than those from mid-tide, which are in turn more resistant than low tide or sublittoral forms. As already noted (p. 86) some of these differences may be due to the thicker cell walls found in the high-level plants, but there may also be associated protoplasmic resistance.

Excessive immersion in the sea can have adverse effects on intertidal plants. Although the form of *Fucus vesiculosus* found in the Baltic appears to thrive at one or two metres depth, some of the high water and mid-tide fucoids from tidal shores cannot survive continuous immersion at as little as one metre depth, and one species has been shown to die if it is kept under illumination reduced to one quarter of that normally experienced. These effects are related in some way to the spectral composition of the light, as a series of experiments on photosynthesis at different depths has demonstrated. Most fucoids and an intertidal laminarian gave the maximum rate of photosynthesis at one metre depth, declining at deeper levels, whereas the optimum for some species of sublittoral red algae was as much as 15 metres depth. The red seaweeds appear to be adapted for life at reduced levels of illumination, largely through the action of their dominant pigment, phycoerythrin, which masks the other pigments (chlorophylls and carotenes) also present. Recent researches indicate that all the pigment molecules are closely associated in the plastids where photosynthesis occurs, and that the phycoerythrin is able to absorb energy in the blue-green and blue wave-lengths that penetrate best in sea water, and then transfer the energy (apparently by fluorescence at a redder wave-length) to the chlorophyll.

Experiments on the rate of photosynthesis can be carried out quite easily. A portion of the plant under test, or a whole plant if small, is carefully washed and any epiphytes removed. It is placed in a tightly closed vessel of sea water (e.g. a stoppered jar with the ground glass surfaces sealed with a trace of vaseline) which is carefully closed to exclude any trace of air bubbles. The vessel is maintained at constant temperature in a water bath (or is lowered into the sea from a pier, boat or buoy to a known depth) and kept under constant illumination (e.g. daylight fluorescent tube) or the illumination in situ is measured with a photometer. A sample of the water in the vessel is taken before sealing, and again on opening after the experiment, and the oxygen content measured by the Winkler method (see Chapter 8). A similar experiment at the same time,

THE CAUSES OF ZONATION 117

but with the experimental vessel placed in a light-tight bag, will show the amount of oxygen consumed in the dark by plants and the respiratory uptake of animals. The results are usually expressed as cc O_2 per gram weight (dry or wet) per hour at a given temperature under given illumination.

Emersion and temperature

If allowance is made for species of different geographical distribution, there is a definite tendency for animals that live at the highest levels on the shore to tolerate higher temperatures and greater desiccation than those at mid-tide or lower levels. There may also be a difference in resistance to these factors between individuals of the same species taken from different tide levels. Further differences of a similar sort exist in the rate of growth, which is much slower in animals that are out of the water for long periods. In many species there appears to be a corresponding difference in some physiological function; for example movement of respiratory appendages, heart beat, and oxygen uptake, which may all have a lower rate at a given temperature in high-level forms, whose metabolic rate is thus less than that of low-tide individuals.

Selection of habitat

In theory, mobile animals should be better able to select the best tide level to suit their metabolism, and might be expected to be critically adapted to their particular zone. However, the experiments mentioned above show that this is only partly true for forms such as the littorinids, top-shells and limpets, and an explanation may lie in the behaviour of these animals. Thus although marked specimens of the common winkle *Littorina littorea* remain in the same area for weeks on end, the individual animals have been shown to move about a good deal. They are particularly active at periods when the tide is falling or rising, and move in U-shaped paths away from their base and back again, apparently employing the direction of the light or sunlight, or even the plane of polarisation in the sky, as a compass in the same way as many terrestrial insects. Whatever the means of navigation employed, the winkles do range over a wider area than that in which each is found at low tide, and hence a greater variation of environment is experienced than might be expected. In their behaviour the winkles clearly show a stage in development of the homing response which is well shown by limpets. The latter have definite homes or scars on the rocks, where each lives for long periods, returning after making feeding excursions.

It is quite simple to demonstrate the attachment of limpets to their homes using one of the groups that are often found near HWN on a rocky shore. Each shell is scraped with a knife, rubbed with a pad moistened with alcohol, and then painted with a quick-drying enamel and the whole group photographed; alternatively the shell can be marked with an engraving tool or dental drill powered from a battery. It is as well to choose a fairly lonely place, and to make the marks as inconspicuous as possible commensurate with finding them again, in order to reduce interference with the experiment. Visits at intervals will show that all or nearly all of the limpets remain in the same area for months or years, and many of them return to the same scar each day as the tide falls. The homing response is probably a form perception of the surrounding rocks through the limpet's tactile senses (e.g. foot and tentacles) but it would not be at all surprising if present and future researches showed the existence of more sophisticated senses and navigation similar to those suspected to occur in winkles. Experiments on this topic would be very rewarding, and might be started by changing the immediate surroundings of a marked group of limpets, for example by chiselling away a lot of the rock surface, and seeing what change took place in the proportion returning to base.

Sharply zoned species

These examples of what might be called dynamic zonation of adult molluscs are of no great help in studying the causes of

ON THE SEA-SHORE

larvae during settlement are probably
these factors operating on the adult,
forms, many of which do not settle
numbers outside the zone of the adu
(p. 106) is of the greatest importance
sting zonation of a species.
ation is primarily caused by the tides a
fied by the differing degrees of tolerance
ersion shown by the organisms, and the
one another, competition, predation an
, which all tend to sharpen the boundarie

Erratum—page 128

Third line from bottom of page:
for '20 cm.' read '10 cm.'

PLATE VII. *Top*, a beach of coarse sand, with slate rock outcrops, on the western coast; *bottom*, another coarse sand beach, showing ripple marks and worm-casts near low water. Only the tail-casts of *Arenicola marina* can be seen clearly here, as the sand has collapsed and partly filled the head shafts to the burrows.

PLATE VIII. *Top*, a sheltered muddy shore in a large estuary. Around mid-tide level the mud is covered with patches of green weed, *Enteromorpha*, and clumps of *Fucus vesiculosus*, and there are also some groups of edible winkles, *Littorina littorea*. Towards low water the mud is much softer, and only the casts of *Arenicola* can be seen; *bottom*, a mud shore, showing well-formed head shafts to the burrows of *Arenicola*, and the castings built up into mounds. Smaller burrows indicate the presence of some crustacea and other worms, and there are some tracks left by crawling winkles.

8

Methods of Studying the Sea-shore

ONLY a few of the practical details of work on the shore could be included in the previous chapters. Now that all the habitats have been described, and the effects of environmental factors discussed we can look at the best means of sampling the animals and plants and of measuring the external factors that influence them.

SURVEYING AND LEVELLING

The first essential is to obtain an adequate map of the area being worked, to the largest possible scale. The 6-inch and 25-inch sheets of the Ordnance Survey are ideal if up to date, but only older surveys may be available for some districts. In these cases the local authority, local river board or harbour commissioners may have their own survey showing coastal defences, sea walls, new roads and so forth, and may allow it to be copied. Whatever map is available it is almost certain that details of the intertidal zone are shown only approximately and that the Ordnance Survey's indication of high-water mark and low-water mark may have little reference to those observed. If, therefore, a full-scale quantitative survey is planned, the area to be investigated should be remapped, using the existing map as a basis. This is best done by a combination of levelling and plane table survey.

For work in the intertidal zone, levels or heights relative to the land and the tides are more important than exact positioning of topographical features, and the best levelling instrument available within the budget should be obtained. Rough work on a sandy beach can be done with a simple sighting device fitted with a spirit level, or with a mining engineer's hand level,[1] but on rocky shores a tripod mounted quick-setting or dumpy level is best. There is sometimes available on the government surplus market a gunnery 'Director', which combines some basic features of a level, a transit theodolite and a bearing compass. This can be fitted onto a camera tripod with slight modifications, and makes an ideal instrument for shore surveys. With this type of instrument it is possible to set up lines of marks by measuring angles from a fixed point on the land shown on the original map, and then proceed to level the marks and calculate their heights. The lines of marks provide a grid on which a plane table survey can be run, and then contours can be plotted.

An easier method of working if time is short is the often used transect or traverse down the shore from high water to low water. This is **a** single line of marks, for example dabs of paint, lined up by a compass bearing. It is levelled by plotting the position of the water's edge at hourly intervals during one tidal period, and working out the corrected height of tide from the predicted high waters and low waters by a method described in the Admiralty tide tables (see Further Reading).

Levelling

The method of using a level is shown in Fig. 50. The instrument is set up steady on the tripod and then levelled approximately, using the spirit level fitted to the base of the instrument and adjusting either 3 or 4 screws (standard level) or a ball and socket device with locking screw ring (quick setting level), or by moving the tripod legs and altering their length (inexpensive model). When the base on which the telescope swivels is approximately level, the telescope can be swung to sight the levelling rod. The latter is usually a set of telescopic sections marked in feet and hundredths (or in a metric scale), and capable of being easily read from a distance. A home made set of poles, e.g. tent poles painted, is quite satisfactory for many

[1] e.g. 'Abney' clinometer. A Japanese version is available quite cheaply from some surplus and instrument dealers.

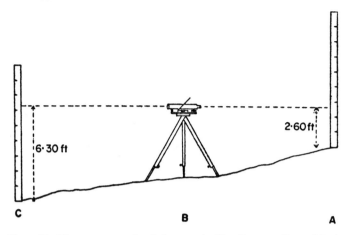

FIG. 50. How to use a level (see text). The first reading with the rod at A is 2·60 feet; the rod is moved to C, the level swung round and re-adjusted to give a second reading of 6·30 feet. Therefore C is 6·30 − 2·60 feet (3·70 feet) below A.

purposes; the important things are to have the marks accurate and legible, and the base of the pole shod with a heavy brass plate to prevent it sinking into the ground or wearing away. After lining up the telescope on the pole, adjustments are made with the main spirit level alongside the telescope and the fine screw, until the telescope tube is dead level. The height visible on the rod coinciding with the centre cross hairs in the telescope is then noted. The rod or pole is now moved to another position, the telescope swung round to sight it, and the fine screw readjusted if necessary to bring the telescope tube dead level again. The difference between the two readings gives the difference in height between the two points on which the rod was placed. If one height is exactly known (e.g. from a previous survey or from the Ordnance Survey bench-marks) then the actual height of the other above sea-level is obtained. Usually a whole sequence of readings is made, starting from a known height and working round back to it in a circle, moving rod and level alternately. The error that is almost certain to be present can be determined by the difference between the first and last readings, and then spread out over all the readings taken. (See Further Reading.)

Plane table

A plane table consists of a small drawing board fitted onto a tripod. The board is set up horizontally by means of a spirit level. A simple sighting device called an alidade is placed on the surface of the board after pinning a sheet of paper on it. The alidade is a stout ruler with graduations along the edge, fitted with sights at each end. Topographic features and limits are plotted by sighting along the alidade and drawing a line parallel to the rule and thus perpendicular to the feature on the paper. The distance to the object is then paced off or taped, and measured off on the line drawn (to an appropriately reduced scale). After most features have been plotted in this way and labelled the points can be connected up freehand. Alternatively, features and objects are sighted from one place, without measuring, and then the table is moved to another position on the other side of the area being surveyed and a new set of sight lines drawn for exactly the same features. If the distance between the two positions of the table (base line) is known the map can be completed freehand by joining together the points where each pair of lines intersects, and the distances found by scale measurement.

Tide readings

Some useful results can be got from 24 hours' observations of the tide at one place if simultaneous observations are made at another place for which full details of the tides are published in tide tables, or are otherwise known. The observers should choose a suitable place, preferably in the shelter of a pier or break-water, and if a tide pole is not already fitted, make one up from a length of wood. This should have a scale marked on it, with alternate feet painted in contrasting colours, subdivided

METHODS OF STUDYING THE SEA-SHORE 125

into 6 inches or 3 inches (or 1/10 feet if the harbour is very calm). Metres and cm. can be used if desired but this makes comparisons difficult as British tide tables are always in feet. The observers synchronise their watches with GMT and then take readings of the height of the water against the pole at each place, at hourly or half-hourly intervals. The scales on the poles need not be arranged in any particular way as long as the relative height of the zero is known, e.g. its distance below the land survey datum, obtained by levelling from a bench mark. The difference in range of tide (ratio) and time can then be applied to the tide data and predictions for the reference place to produce an approximate tide table for the place being observed.

Observations can be made on the open coast in calm summer weather, by erecting a rough tide pole in a cleft in the rocks and tying it down with wire fastened to large nails or specialised fastening devices driven into the rock.

QUANTITATIVE SAMPLING

Sand and mud shores

The best method is probably to mark out a series of stations at equal intervals (e.g. at 1 or 2 feet intervals vertically by the tide; or at 10, 20, 50 or 100 feet intervals horizontally by pacing or taping) along a simple line traverse. At the chosen points a wire or steel frame, of 1/4 or 1/10m.^{2} area, is laid on the deposit or pressed into the surface, and the sand or mud within it rapidly dug out to a spade's depth and placed in a large plastic or galvanised iron bath (e.g. 'baby bath'); more than one may be needed. The deposit below the first spit is then forked over quickly to see if any *Arenicola*, *Nephthys* or large molluscs have been missed. The bathful of deposit is taken down to the water's edge and washed through a sieve, a little at a time. The sieve should be a wooden box frame 12 to 24 inches long by 12 to 18 inches wide and up to 12 inches deep, the bottom covered with a sheet of perforated zinc or woven brass mesh with holes of 1 or 2 mm. diameter. The deposit is washed through by dunking and stirring, and if necessary the material can be puddled in the bath before adding to the sieve. Large stones should be picked out and then the whole contents of the sieve washed out into a glass jar. If the contents are not to be examined straight away they should immediately be preserved (e.g. 5 to 10% of commercial formalin in sea water—see p. 132) and labelled. For the smaller fauna and larval forms a sample of 1/100 m.² might be dug out and passed through a sieve with holes of 0·2 to 0·5 mm. diameter.

A whole beach can be covered by making several line traverses, and it will then be possible to show the abundance of the animals on contoured maps.

Interstitial fauna

Various complicated machines have been described for obtaining subsoil samples of the interstitial microfauna, but for most purposes on the beach two methods will be sufficient. For the surface forms found in moist intertidal sands it is enough to scoop out small quantities shallowly into a 1 litre cylinder until 100 cc. is obtained. The cylinder is then filled with sea water that has been passed through a very fine piece of netting (200 meshes per inch or finer) and thoroughly shaken up. The water is then poured off through a net of the same mesh, and the catch on the net placed in a jar for examination alive the same day. A variant of this method is to take the sand back to the laboratory and treat it with a simple narcotic (e.g. a 7% solution of magnesium chloride in distilled water). All clinging forms will relax and can be filtered off after stirring up.

A second method is useful for the subsoil fauna of a beach much above the tideline or on tideless shores. A hole is dug to about 10 or 20 cm. deep and all the topsoil removed. When the subsoil water begins to drain into the hole the lower layers of deposit can be puddled, and the animals obtained by repeatedly scooping the water with a fine net.

Rocky shores

If only a simple traverse is being investigated it can be marked out with dabs of paint. Then, after levelling and working out the heights, a series of stations can be chosen at fixed intervals and marked with contrasting dabs. If an area is being fully surveyed, the marks are better placed more permanently; quick-setting cement is mixed with sea water and used to fill in natural holes in the rock or laid on the surface after keying it with a chisel. The cement should be marked with letters or figures before it sets, and will last for several years even on wave-swept shores. If the rock is soft enough it can be marked directly with a chisel.

At each station counts are made within a wire frame laid down on the rock; for large animals (limpets) this should be 1 m.^{2}, but 1 dm.^{2} or less will serve for barnacles and littorinids. The algae can be estimated as % cover of the rock by using a 1-m.^{2} frame divided up into one hundred 1-dm. squares or a more random method can be used. In the random method a 1/10-m.^{2} frame is cast over the shoulder 10 times at each station, and the occurrence of an algal species recorded as scores from 1 to 10. In addition to the fixed stations the upper and lower limits of the organisms should be plotted by measuring or by means of a plane table. Samples of the algae and animals should be removed from a definite area and weighed, and small subsamples should be preserved for the checking of identifications.

Methods such as these can be applied only to fairly smooth sloping shores. Where the rock is much broken a qualitative survey is all that can easily be undertaken, although a flexible type of metre grid has been suggested for the purpose. Alternatively, small areas might be counted for the smaller organisms only. With shores containing both rocky and sandy areas, surveys have to be improvised as best possible.

Analysis of sands and muds

When sampling the fauna several lots of the deposit should be scooped up as digging proceeds, until enough is taken to fill a 1-lb. screw-topped jar. Alternatively, if it is wished to analyse the undisturbed material, a simple core sampler can easily be constructed from a length of 5 cm. diameter brass piping. One end is sharpened with a file, the other end is provided with a strengthening collar through which two holes are drilled to take a cross handle. With two people thrusting down on the handle and twisting at the same time the tube can easily be pushed one or two spade depths below the surface, or until stopped by gravel or stones. The core remains inside the tube when it is pulled out of the beach and must be pushed out with a rammer (e.g. a broomstick provided with a brass head). Each 5 cm. length is cut off as it emerges from the corer and placed in a labelled jar. A second similar sample might be taken at the same time for analysis of the interstitial microfauna.

Each sample of sand or mud should be well stirred up and a smaller lot removed from it, weighed wet and dried to constant weight at 105° C. to determine the moisture content. Another sub-sample might be taken for analysis of organic matter or interstitial water (see below) and the remainder dried and stored.

For the standard method of mechanical analysis, about 20–25 g. of the dried deposit is shaken up with a 10% solution of sodium carbonate until all lumps disappear and all silt seems to be in suspension. The shaking is usually performed with a motor-driven device in agricultural practice, but this may not be necessary with shore materials. The shaken deposit is next washed through a 2-mm. round-holed sieve, with the aid of a wash bottle. The material left on the sieve is dried and weighed and constitutes the *gravel*. The remainder passing through the first sieve is then washed through a 0·2-mm. mesh. The material retained is dried and weighed as *coarse sand*. The material passing the second sieve is placed in a large beaker and made up with distilled water to a height of 20 cm. from the bottom of the beaker. Preferably this beaker should stand in a larger vessel of water maintained at 20° C., particularly if the room is draughty. After the liquid has stood undisturbed for 5 minutes the supernatant liquid is poured off, the beaker refilled and stirred thoroughly, and the process repeated until the supernatant liquid looks clear.

The *fine sand* fraction remains in the beaker and is dried and weighed. The *silt* and *clay* that have been poured off can be collected, dried and weighed also, or else can be estimated as the difference between the original weight of the sample and the sum of the other three fractions. Determination of silt and clay separately is more difficult and requires a temperature-controlled room free from draughts, or a thermostatic bath. The suspension remaining after the coarse sand has been separated is thoroughly shaken up in a 1-litre cylinder, after making up to 1 litre, and allowed to settle for 4 minutes 48 seconds at 20° C. A 20 ml. pipette is then carefully lowered into the column of water by rack and pinion, until 10 cm. below the surface, and filled by steady suction. The sample in the pipette is dried and weighed and gives an estimate of the *silt* and *clay* present in the whole column of water. A similar sample taken at 10 cm. depth 8 hours after shaking, also at 20° C., will give an estimate of the *clay* fraction, after which the fine sand is obtained by repeated shaking up and pouring off as in the simpler method already described.

Interstitial water

Water held between the particles of a sample of sand or mud can usually be removed only by suction through filter paper on a Buchner funnel. The resulting liquid is available for analysis of salinity and for pH tests.

Organic matter

In conventional agricultural analysis of soils it is usual to remove the organic matter before submitting a sample to mechanical analysis. Originally this was carried out by incineration in a closed crucible, and the difference in weight before and after regarded as an approximate index to organic content. Unfortunately, particularly in beach sands, this method also drives off much carbon dioxide from carbonate that is present, as for example in fragments of shell, and is now regarded as unreliable. A later method involves heating the sample for a fixed time with a strong solution of hydrogen peroxide. This method is also only approximate and is subject to the deficiencies of all methods of organic matter analysis that are based on oxidising agents. With shore samples the organic content is generally low enough not to introduce significant error into the type of mechanical analysis advised.

If it is wished to analyse the organic matter then recourse should be made to one of the quantitative methods involving digestion of the sample with hot acidified dichromate or permanganate, followed by titration of the excess oxidising agent by standard volumetric methods. The easiest method, and one that has given consistent results in the author's hands is that of Schollenberger modified by Allison, and later by Anderson.

A solution of potassium dichromate is made up containing 19·61 g./l. 10-ml. pipette samples of this are placed in 150×25-mm. Pyrex test tubes and evaporated down to dryness at 80 to 85° C., afterwards being stored in a desiccator. About half a gram of the dried sand or mud is weighed out into one of the test tubes containing dichromate, followed by 10 ml. concentrated sulphuric acid and 0·2 g. silver sulphate. The tube is heated steadily over a low bunsen flame (2 to 3 cm.), stirring constantly to the bottom with a 350° C. thermometer, the heating being adjusted so as to reach 175° C. in $1\frac{1}{2}$ minutes. (Avoid heating too rapidly, or above 180° C., otherwise the acid will fume.) The tube is at once removed from the flame, allowed to cool 5 minutes in air and under the tap. When cool the contents are poured out into 50 ml. of distilled water, the tube rinsed out, and the final volume adjusted to 150 ml. About 5 g. of sodium fluoride is added and the mixture titrated against a standardised solution of ferrous ammonium sulphate (78·6 g./l. containing 20 ml. concentrated sulphuric acid) using 3 drops of diphenylamine as indicator (change at end point from blue to green). A blank determination without a sample of deposit is made with each series and the value deducted from the burette reading obtained with the sample. Theoretically, 1 ml. of the ferrous ammonium sulphate solution is equivalent to 0·12 g. organic carbon. However, by comparison with standard combustion train gravimetric methods, and from tests made with quantities of known compounds, recovery is not quite 100% and the practical value is 1 ml. ferrous ammonium sulphate to 0·138 g. organic carbon.

A more recent version of this technique gets rid of the salt in the sand or mud sample by washing on a sintered glass funnel or hard fine filter paper. The sample is then dried and passed through a 0·5-mm. sieve. About 0·15 to 0·3 g. is weighed into a 15 × 250-mm. Pyrex tube, and 10 ml. chromic acid added from a wide-tipped pipette. A loose bulb stopper is fitted and the tube heated in a boiling-water bath for 15 minutes. The tube is cooled under the tap and the contents poured into 200 ml. water, and titrated against ferrous ammonium sulphate solution, using one drop ferrous phenanthroline indicator. The end point is a clearly marked change to a persistent pink colour. A blank titration, without sand or mud, is carried out and the result deducted from the final value. The reagents are chromic acid: 13 g. CrO_3 is dissolved in the smallest possible amount of water, then 900 ml. of concentrated H_2SO_4 is added and the mixture made up to 1 litre when cold; ferrous ammonium sulphate: 0·2 N containing 2% H_2SO_4; ferrous phenanthroline: 0·337 g. of o-phenanthroline monohydrate in 25 ml. of 0·695% ferrous sulphate. With these concentrations 1 ml. of the ferrous ammonium sulphate is equivalent to 1·15 × 0·6 mg. organic carbon, as corrected by combustion train determinations.

In another method devised for plankton, the salts present in the sample are not removed, but the chloride is volatilised as free chlorine by pretreatment with orthophosphoric acid.

These methods, like all other oxidising techniques, attack relatively inert materials such as coal and coke, and do not of course distinguish between utilisable organic matter and materials such as cellulose and faecal products. For such reasons they cannot be used on beach materials until these have been examined microscopically for coal and coke particles. Even if the sample appears free of such materials, the result can still only be regarded as approximate.

Preserving samples

Commercial formalin (approximately 40% HCHO in water) is often very acid, due to the presence of formic acid; it becomes very cloudy with age and deposits a hard white sediment. If used directly to make up preserving solutions the acid may dissolve shells and skeletons, and the turbidity can prevent us seeing what the jar contains. It is best therefore to neutralise the formalin and filter before diluting to the usual working strength of 1 part to 9 parts of sea water.

The simplest and best method is to add 25 g./l. of Hexamine (hexamethylene tetramine) to the formalin and then filter through filter paper or well-packed glass wool. The Hexamine is an excellent buffer and will keep the formalin neutral for months. A more awkward but cheaper operation is to add small doses of calcium hydroxide until neutrality is reached, and then an excess of calcium carbonate, which is allowed to remain in the jar. The liquid has to be filtered each time some is needed, or else decanted off very carefully.

Some workers use formalin only in the field and then transfer the sample to 70% ethyl alcohol or to iso-propyl alcohol for storage. Care should be taken that the alcohol itself is not acid.

For prolonged storage formalin can be replaced (it must be used first, to fix the tissues) with solutions of 1% propylene phenoxetol or 0·1% Nipa ester in distilled water.

OTHER ENVIRONMENTAL FACTORS

Temperature

Any reasonably accurate mercury thermometer can be used to take the temperature of the sea, tide-pools and sand. It is best to obtain one with graduations of 0·1 or 0·2°, and to check it against a certificated thermometer. For use in sand the instrument should be calibrated for a few centimetres immersion so that the mercury can be read while the bulb is within the deposit. By digging a hole with a spade and pushing the thermometer into the side, readings can easily be taken at various depths.

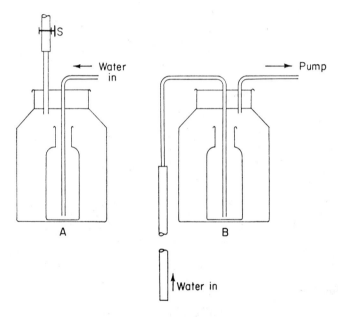

FIG. 51. Simple device for taking water samples at the bottom of estuaries or tide pools, for salinity and oxygen analysis. A is lowered to the bottom with the water flow adjusted by partly closing screw-cock S, so that the inner bottle is completely flushed through after it reaches the bottom. B, is kept at the surface, the rubber or plastic inlet tube being long enough to reach the bottom. The pump suction needed depends how far the bottle is above the surface of the water.

For sea water, the thermometer is best mounted in a frame of wood, provided with a little bucket on the end so that a sample of the water covers the bulb when it is withdrawn from the sea. If possible, 4 or 5 minutes should be allowed for the frame and thermometer to equilibrate with the water before taking them out to read the temperature.

To measure microclimates close to the ground, and the body temperature of animals, it is best to use thermocouples or thermistors. The former are smaller and easier to construct, but require an accurate portable galvanometer or potentiometer and an insulated reference junction. Thermistors must be purchased, and are slightly more bulky, which may be a disadvantage with small animals, but on the other hand they require only a simple bridge circuit and small milliammeter. Those skilled enough to make up thermocouple or thermistor units will find appropriate references in the section on Further Reading.

To sample deep water in the sea or water at the bottom of an estuary, special sampling bottles are needed. There are many complicated types in use for oceanography, and one of these has been adapted for use in shallow waters, on the end of a rope, but all are expensive and unlikely to be readily available.

A simpler arrangement consists of a standard reagent bottle inside a weighted can or larger bottle (Fig. 51A). Alternatively, a length of plastic tubing with a weight at the end is lowered to the required depth and water is drawn into the sample bottle by suction with an air pump (Fig. 51B).

Humidity

Readers may be familiar with wet- and dry-bulb thermometers used to measure atmospheric humidity for meteorological purposes. A similar instrument, in a swivel frame (whirling hygrometer), can be used to measure the humidity of the air over the shore. To measure the humidity close to the rocks needs small paper or hair hygrometers which can be purchased from dealers. Electronic methods involving sensing devices that measure either water content of the air directly or change in temperature of evaporating liquid films, are most accurate but are rarely needed on the sea-shore (see Further Reading).

Salinity

There are two comparatively easy ways to measure salinity. For estuaries, or where the fluctuation is large, it may be enough to measure the specific gravity of the sample with a large, accurately calibrated hydrometer. These can be obtained ready marked in salinity values, but require correcting for the temperature of the sample. The scientific supply houses will sell the hydrometers (usually three are needed to cover the whole range from 0 to 35‰) together with the necessary correction table.

Where accuracy is deemed more important, as on some shores and in the higher reaches of estuaries where salinity is low, it is best to estimate the chlorinity of the sample by titration and then to convert the titration values to salinity by the use of tables (see Further Reading).

For most purposes it is sufficient to titrate a 10-ml. sample of the water against a solution of silver nitrate containing 27·25 g./l. (this should be stored in the dark). The end point of the titration is determined with a few drops of 10% potassium chromate solution as indicator, the change from bright yellow to deep orange being quite distinct. The volume of silver nitrate solution in ml. used is equivalent to the salinity of the sample in parts per thousand subject to minor correction shown on next page. The correction is needed because salinity is a wt./wt. relationship, not a volumetric concept.

A similar technique using more accurate semi-automatic pipettes and burettes was once the standard procedure for oceanographic analysis of salinity, often with the end point determined by a valve electrometer. Nowadays it has become customary to use a very accurate conductivity cell under perfectly controlled temperature conditions, thus allowing a much larger number of samples to be analysed in a given time.

Simpler devices, using induction principles, with built-in temperature compensators, are now in use for oceanographic work and may eventually become more widely employed.

salinity found ‰	correction	salinity found ‰	correction
40	−0·15	24	+0·20
38	−0·08	22	+0·22
36	−0·03	20	+0·23
34	+0·03	18	+0·23
32	+0·07	16	+0·23
30	+0·11	14	+0·20
28	+0·15	12	+0·19
26	+0·17	10	+0·16

Hydrogen ion concentration

On the shore a glass electrode pH meter is of doubtful value owing to the limited temperature range available. It is possible to use one for interstitial water, which can be removed by drainage or suction from a sample of sand in the laboratory. In the field the best method is still a set of buffer solutions with a suitable indicator. The buffer solutions must be made up to the scale of values needed, and the boric acid-borate mixtures are easiest (e.g. Palitzsch). Only two accurate microburettes are necessary to make up a series of mixtures from two standard volumetric solutions. For some purposes a set of 'Capillators' may be quite useful, but pH test papers are less so, owing to the limited range of pH values found in the marine environment.

Oxygen content

Descriptions of the Winkler method for analysis of oxygen content of water samples will be found in the standard textbooks of water analysis (see Further Reading).

For work on respiration and photosynthesis, micromethods are preferable. Several modifications of the Winkler method have been described (see Further Reading). Apparatus for this purpose can be bought ready made from the manufacturers,[1] who will also supply full instructions.

A more recent rapid microtechnique employs a polarographic diffusion method, but at the moment it needs careful control and comparison with Winkler analyses, and is more suitable for long-term experimental studies.

Aquaria

Some observations on habits of shore animals can be made with simple marine aquaria. The jars or tanks should be made of materials that are not corroded by sea water—glass, stainless steel, slate and the harder plastics are all suitable. The essential requirement is some clean sea water of high salinity (over 30‰). This can be brought back from an open-coast shore in well scrubbed-out glass or polythene carboys; or it can be obtained for the cost of carriage and a nominal charge from some marine laboratories.[2] If all else fails some of the hardier animals will exist for a while in artificial sea water (Table 7). The formula has been devised by chemists and physiologists for studying the effect of variation in ionic composition or trace elements on the behaviour of marine organisms.

The tanks should be well cleaned out and given a thin bottom layer of pebbles or sand that has been washed in several changes of fresh water and then in sea water. After the sea water is added it must be kept continuously agitated and aerated with a small pump or compressed air. Sea water containing organic matter goes bad much more quickly than fresh water, and under reducing conditions hydrogen sulphide is formed from the sulphate present. Only a few animals should be put into the tank at first and all those that die must be removed as soon as possible. Any flocculent organic matter gathering at the bottom must be removed with a large bulb pipette or a siphon.

[1] G. T. Harris & Co., Birmingham.
[2] e.g. Marine Biological Association, Citadel Hill, Plymouth.

Table 7

Artificial Sea Water

The following compounds of A. R. quality, weighed out and made up to one litre with distilled water, will produce a passable imitation of sea water. For many short-term experiments the last four ingredients can be omitted. When dissolved, the solution should be well aerated by bubbling air through it with a compressor for about half an hour, and then the pH adjusted, if necessary, to 8·0–8·4, by adding very small quantities of normal solutions of HCl or NaOH.

	g/l.
NaCl	24·087
KCl	0·681
$CaCl_2$	1·131
$MgCl_2$	5·111
Na_2SO_4	4·019
$NaHCO_3$	0·197
KBr	0·098
H_3BO_3	0·027
$SrCl_2$	0·025
NaF	0·003

Experience will show which animals can be kept under these conditions, but sea-anemones and mussels always seem to do well, and further experiments should be made with the animals from the higher tide pools. Barnacles may survive but are not as easy to feed as the anemones which will take small pieces of meat or fish. Molluscs such as limpets and winkles are less suitable as they tend to come out of the water and wander around. They may travel so far that they get lost or dry up and die. On the other hand they are very useful in keeping the glass clear of encrusting algae, on which they can be seen grazing. However, it is best to screen the tanks from daylight, as the weed growth may get out of hand.

Food for some of the smaller animals can be provided by

rearing the brine shrimp, *Artemia salina*, from the eggs sold by dealers. The eggs are sprinkled on the surface of the water in a separate tank; the larvae that hatch out can be fed on a suspension of dried yeast in sea water, added in small quantities at frequent intervals. With luck the animals can be reared to full size and will then start breeding so that fresh cultures can be maintained by removing the newly hatched larvae that result. The larvae and adults provide food for several different sizes of animals.

The engraving below shows a Victorian naturalist investigating a rocky shore, complete with silk hat, and followed by an assistant to carry a shrimp-net and other collecting gear.

9

Further Reading

THERE are several more general books that can be read with advantage in addition to more advanced works dealing with special topics. However, anyone wishing to delve further will, at some stage, have to consult original papers, particularly those detailing results of more recent work, which has been noted only briefly in preparing this short account. Some of the journals in which results of marine research are published may be found only in University libraries, Marine Laboratories, and some of the larger public reference libraries. However, it should be possible to consult these at a local reference library by means of the inter-library loan system.

Of course, only a small selection of original papers can be given here, but those listed will be found to have extensive bibliographies of publications on the same and related subjects.

General

ANDREWARTHA, H. G. & BIRCH, L. C. (1954). *The Distribution and Abundance of Animals*. Chicago University Press.

BARRETT, J. H. & YONGE, C. M. (1958). *Collins Pocket Guide to the Sea-Shore*. Collins.

EALES, N. B. (1950). *The Littoral Fauna of Great Britain*. Cambridge University Press.

EKMAN, S. (1953). *Zoogeography of the Sea*. Sidgwick & Jackson.

HEILBRUNN, L. V. (1943). *An Outline of General Physiology*. Saunders.

HEDGPETH, J. W. (Editor). (1957). *Treatise on Marine Ecology and Paleoecology*, Vol. I. Geol. Soc. America, Mem. 67.

LEWIS, J. (1964). *The Ecology of Rocky Shores*. English Universities Press.

NICOL, J. A. C. (1960). *The Biology of Marine Animals*. Pitman.

FURTHER READING

PEARSE, A. S. (1950). *The Emigrations of Animals from the Sea*. Sherwood Press.
PROSSER, C. L. & BROWN, F. A. (1961). *Comparative Animal Physiology*. 2nd edition. Saunders.
RICKETTS, E. F. & CALVIN, J. (1962). *Between Pacific Tides*. 3rd edition, revised by J. W. Hedgpeth. Stanford University Press.
SMITH, J. E. (1953). Maintenance and spread of sea-shore faunas. *Advanc. Sci., Lond.*, 10 (No. 38), pp. 145–156.
WILSON, D. P. (1935). *Life of the Shore and Shallow Sea*. Nicolson & Watson.
YONGE, C. M. (1949). *The Sea-Shore*. Collins.

Chapter 1

ALEXANDER, W. B. (1955). *Birds of the Ocean*. Putnam.
JENKINS, J. T. (1936). *Fishes of the British Isles*. Warne.
MARINE BIOLOGICAL ASSOCIATION OF THE U.K. (1957). *Plymouth Marine Fauna*. 3rd edition.
PARKE, M., & DIXON, P. S. (1964). A revised check-list of British marine algae. *J. mar. biol. Ass. U.K.*, 44, pp. 499–542.
SCHEFFER, V. B. (1958). *Seals, Sea Lions and Walruses*. Stanford University Press.
TINBERGEN, N. (1951). *The Study of Instinct*. Oxford University Press.

Chapter 2

ADMIRALTY, HYDROGRAPHIC DEPARTMENT (published each year). *Admiralty Tide Tables*, Vol. I, *European Waters*.
DOODSON, A. T. & WARBURG, H. D. (1941). *Admiralty Manual of Tides*. H.M.S.O., London.
GUILCHER, A. (1958). *Coastal and Submarine Morphology*. Methuen.
HARVEY, H. W. (1955). *The Chemistry and Fertility of Sea-Waters*. Cambridge University Press.
HILL, M. N. (Editor) (1963). *The Sea*, Vol. 2, *The Composition of Sea Water*. Interscience.
KING, C. A. M. (1961). *Beaches and Coasts*. Edward Arnold.
RUSSELL, R. C. H. & MACMILLAN, D. H. (1952). *Waves and Tides*. Hutchinson's.

STEERS, J. A. (1946). *The Coastline of England and Wales*. Cambridge University Press.
STEERS, J. A. (1953). *The Sea Coast*. Collins.
STEERS, J. A. (Editor) (1960). *Scolt Head Island*, (revised edition). Heffer.
VINOGRADOV, A. P. (1953). *The Elementary Chemical Composition of Marine Organisms*. Sears Foundation, Mem. 2.

Chapter 3

CHAPMAN, V. J. (1957). Marine algal ecology. *Bot. Rev.*, 23, pp. 320–350.
COLMAN, J. (1933). The nature of the intertidal zonation of plants and animals. *J. mar. biol. Ass. U.K.*, 18, pp. 435–476.
EVANS, R. G. (1947). The intertidal ecology of selected localities in the Plymouth neighbourhood. *J. mar. biol. Ass. U.K.*, 27, pp. 173–218.
FELDMANN, J. (1951). Ecology of marine algae. In: *Manual of Phycology*. (ed. G. M. SMITH), pp. 313–334. Chronica Botanica.
HUTCHINS, L. W. (1947). The bases for temperature zonation in geographical distribution. *Ecol. Monogr.*, 17, pp. 325–335.
MOKYEVSKY, O. B. (1960). Geographical zonation of marine littoral types. *Limnol. Oceanogr.*, 5, pp. 389–396.
SOUTHWARD, A. J. (1958). The zonation of plants and animals on rocky sea-shores. *Biol. Rev.*, 33, pp. 137–177.
SOUTHWARD, A. J. & CRISP, D. J. (1963). *Catalogue of main marine fouling organisms*. Vol. 1. *Barnacles*. O.E.C.D., Paris.
STEPHENSON, T. A. & STEPHENSON, A. (1949). The universal features of zonation between tide-marks on rocky coasts. *J. Ecol.*, 37, pp. 289–305.

Chapter 4

BRUCE, J. R. (1928). Physical factors on the sandy beach. Part I. Tidal, climatic and edaphic. *J. mar. biol. Ass. U.K.*, 15, pp. 535–552.
BRUCE, J. R. (1928). Physical factors on the sandy beach. Part II. Chemical changes. *J. mar. biol. Ass. U.K.*, 15, pp. 553–565.

CHAPMAN, G. (1949). The thixotropy and dilatancy of a marine soil. *J. mar. biol. Ass. U.K.*, 28, pp. 123–140.

COLMAN, J. S. & SEGROVE, F. (1955). The fauna living in Stoup Beck Sands, Robin Hood's Bay (Yorkshire, North Riding). *J. anim. Ecol.*, 24, pp. 426–444.

DAHL, E. (1952). Some aspects of the ecology and zonation of the fauna on sandy beaches. *Oikos*, 4, pp. 1–27.

HOLME, N. A. (1949). The fauna of sand and mud banks near the mouth of the Exe estuary. *J. mar. biol. Ass. U.K.*, 28, pp. 189–237.

PIRRIE, M. E., BRUCE, J. R. & MOORE, H. B. (1932). A quantitative study of the fauna of the sandy beach at Port Erin. *J. mar. biol. Ass. U.K.*, 28, pp. 279–296.

REES, C. B. (1940). A preliminary study of the ecology of a mud-flat. *J. mar. biol. Ass. U.K.*, 24, pp. 185–199.

SWEDMARK, B. (1964). The interstitial fauna of marine sand. *Biol. Rev.*, 39, pp. 1–42.

WELLS, G. P. (1957). The life of the lugworm. *Penguin Books, New Biology*, 22, pp. 39–55.

WILLIAMSON, D. I. (1951). Studies in the biology of the Talitridae (Crustacea, Amphipoda). *J. mar. biol. Ass. U.K.*, 30, pp. 73–99.

Chapter 5

ALEXANDER, W. B., SOUTHGATE, B. A. & BASSINDALE, R. (1935). Survey of the river Tees, Part II. The estuary—chemical and biological. *Techn. Pap. Water Poll. Res. Lond.*, 5, pp. 1–171.

HYNES, H. B. N. (1954). The ecology of *Gammarus duebeni* Lilljeborg and its occurrence in fresh water in western Britain. *J. Anim. Ecol.*, 23, pp. 38–84.

KINNE, O. (1963). The effects of temperature and salinity on marine and brackish water animals. *Oceanogr. mar. Biol. Ann. Rev.*, 1, pp. 301–340.

SPOONER, G. M. & MOORE, H. B. (1940). The ecology of the Tamar estuary. VI. An account of the macrofauna of the intertidal muds. *J. mar. biol. Ass. U.K.*, 24, pp. 283–329.

Chapter 6

ALLEE, W. C. & OESTING, R. (1934). A critical examination of Winkler's method for determining dissolved oxygen in respiration studies with aquatic animals. *Physiol. Zool.*, 7, pp. 509–541.

BULLOCK, T. H. (1955). Compensation for temperature in the metabolism and activity of poikilotherms. *Biol. Rev.*, 30, pp. 311–342.

CARLISLE, D. B. & KNOWLES, F. G. W. (1959). *Endocrine Control in Crustaceans*. Cambridge Monographs in Experimental Biology, No. 10.

COTT, H. B. (1940). *Adaptive Colouration in Animals*. Methuen.

CRISP, D. J. (1958). The spread of *Elminius modestus* Darwin in north-west Europe. *J. mar. bid. Ass. U.K.*, 37, pp. 483–520.

DAVENPORT, D. (1955). Specificity and behaviour in symbiosis. *Quart. Rev. Biol.*, 30, pp. 29–46.

FOX, H. M. & WINGFIELD, C. A. (1938). A portable apparatus for the determination of oxygen dissolved in a small volume of water. *J. exp. Biol.*, 15, pp. 437–445.

HARKER, J. E. (1958). Diurnal rhythms in the animal kingdom. *Biol. Rev.*, 33, pp. 1–52.

HOAR, W. S. (1953). Control and timing of fish migration. *Biol. Rev.*, 28, pp. 437–452.

HODGMAN, C. D., WEAST, R. C. & SELBY, S. M. (1961). *Handbook of Chemistry and Physics* (43rd edition). Chemical Rubber Publishing Co.

HOWIE, D. I. D. (1959). The breeding of *Arenicola marina*. *J. mar. biol. Ass. U.K.*, 38, pp. 395–406.

JONES, J. D. (1954). Observations on the respiratory physiology and on the haemoglobin of the polychaete genus *Nephthys* with special reference to *N. hombergi* (Aud. & M.-Edw.). *J. exp. Biol.*, 32, pp. 110–125.

JØRGENSEN, C. B. (1955). Quantitative aspects of filter feeding in invertebrates. *Biol. Rev.*, 30, pp. 391–454.

KNIGHT-JONES, E. W. (1955). The gregarious setting reaction of barnacles as a measure of systematic affinity. *Nature, Lond.*, 175, pp. 266.

KNIGHT-JONES, E. W. (1951). Gregariousness and some other aspects of the setting behaviour of *Spirorbis*. *J. mar. biol. Ass. U.K.*, 30, pp. 201–222.

NAYLOR, E. (1958). Spontaneous tidal and diurnal rhythms of locomotory activity in *Carcinus maenas* (L.). *J. exp. Biol.*, 35, pp. 602–610.

SAND, A. (1935). The comparative physiology of colour response in reptiles and fishes. *Biol. Rev.*, 10, pp. 361–382.

THORSON, G. (1950). Reproductive and larval ecology of marine bottom invertebrates. *Biol. Rev.*, 25, pp. 1–45.

WILLIAMS, G. C. (1957). Homing behaviour of California rocky-shore fishes. *Univ. Calif. Publ. Zool.*, 59, pp. 249–284.

YONGE, C. M. (1928). Feeding mechanisms in the invertebrates. *Biol. Rev.*, 3, pp. 21–76.

Chapter 7

BAKER, S. M. (1909). On the causes of the zoning of brown seaweeds on the sea-shore. *New Phytol.*, 8, pp. 196–202.

BIEBL, R. (1952). Ecological and non-environmental constitutional resistance of the protoplasm of marine algae. *J. mar. biol. Ass. U.K.*, 31, pp. 307–315.

KNIGHT-JONES, E. W. & MOYSE, J. (1961). Intraspecific competition in sedentary marine animals. *Symp. Soc. Exp. Biol.*, 15, pp. 72–95.

SMITH, J. E. & NEWELL, G. E. (1955). The dynamics of the zonation of the common periwinkle (*Littorina littorea* (L.)) on a stony beach. *J. anim. Ecol.*, 24, pp. 35–56.

SOUTHWARD, A. J. (1958). Notes on the temperature tolerances of some intertidal animals in relation to environmental temperatures and geographical distribution. *J. mar. biol. Ass. U.K.*, 37, pp. 49–66.

THORPE, W. H. (1963). *Learning and instinct in animals*. Methuen.

ZANEVELD, J. (1937). The littoral zonation of some Fucaceae in relation to desiccation. *J. Ecol.*, 25, pp. 431–468.

Chapter 8

ALLISON, L. E. (1935). Organic soil carbon by reduction of chromic acid. *Soil Sci.*, 40, p. 311.

ANDERSON, D. Q. (1939). Distribution of organic matter in marine sediments and its availability to further decomposition. *J. mar. Res.*, 2, pp. 225–235.

BARNES, H. (1959). *Apparatus and Methods of Oceanography*, Part I *Chemical*. Allen & Unwin.

BOOKER, P. G. (1961). New sea temperature measuring devices. *J. Cons. Int. Explor. Mer*, 26, pp. 133–147.

EDNEY, E. B. (1953). The construction and calibration of an electrical hygrometer suitable for microclimatic measurements. *Bull. ent. Res.*, 44, pp. 333–342.

HYDROGRAPHIC DEPARTMENT, ADMIRALTY (1948). *Admiralty Manual of Hydrographic Surveying* (2nd edition).

KANWISHER, J. (1959). Polarographic oxygen electrode. *Limnol. Oceanogr.*, 4, pp. 210–217.

KNOWLES, F. G. W. (1953). *Fresh water and Salt-water Aquaria.* Harrap.

KØIE M. (1948). A portable alternating current bridge and its use for microclimatic temperature and humidity measurements. *J. Ecol.*, 36, pp. 269–282.

KRUMBEIN, W. C. & PETTIJOHN, F. J. (1938). *Manual of Sedimentary Petrography.* Appleton-Century Crafts, New York.

MARSHALL, S. M. & ORR, A.P. (1964). Carbohydrate and organic matter in suspension in Loch Striven during 1962. *J. mar. biol. Ass. U.K.*, 44, pp. 285–292.

STRICKLAND, J. D. H. & PARSONS, T. R. (1960). A manual of seawater analysis. *Bull. Fish. Res. Bd. Canada*, 125.

VINDEN, J. S. (1961). *The Home Aquarium.* Pan Books.

EL WAKEEL, S. K. & RILEY, J. P. (1957). The determination of organic carbon in marine muds. *J. Cons. Int. Explor. Mer*, 22, pp. 180–183.

Glossary

Bench-mark—a mark made by the Ordnance Survey showing that a point has been accurately levelled during a survey. It is shown on the map as, for example, *B.M.* 16.56, indicating the height in feet above ordnance datum, and reference to the notes at the foot of the sheet will show which datum was in use for the survey (see also Table 3). A bench mark is recognised in the field as a broad arrow topped by a horizontal line, usually carved a foot or two above ground level on vertical surfaces such as gateposts, walls and buildings. To sight from the mark, we can hold a thin sheet of wood or metal in the horizontal groove and place the base of the levelling rod on this.

Buffer—action of certain solutions in opposing change in ionic composition, particularly in the hydrogen ion concentration.

Capillary attraction—in deposits—a manifestation of surface tension, whereby water enters or is held in narrow passages or spaces against the force of gravity.

Declination—angular distance of sun, moon, etc., from a fixed reference point (celestial equator) measured as an arc of a circle polewards.

Discontinuity—of distribution—a place where a species becomes very rare or is no longer found, even though there is a suitable habitat for the particular species.

Dissociation curve—of blood—graph relating oxygen tension in the surrounding air or water to the oxygen content of the blood. Represents the proportion of haemoglobin to oxyhaemoglobin.

Epiphytes—plants or animals that grow on the surface of plants. Similarly, epizooic means growing on the skin, cuticle or shell of animals. Some species always grow in this way, others may grow on rocks and stones as well.

Ionisation—of aqueous solutions—dissociation of molecules into component ions.

Interstices—of sands and muds—the spaces between the grains or particles. Usually contain water (held by capillary attraction) and some air.

Niche—in ecology—status of an animal or plant in its community, particularly its relationships to other organisms and its nutritional habits.

pH—$\log^{10}$ of the reciprocal of the concentration of hydrogen ions, a measure of acidity.

Plankton—floating or drifting life in the sea, both plants and animals, usually not more than a few mm. in size and carried passively in the water currents; those that can swim do so to change or adjust their depth in the water not to move from place to place. In contrast, *nekton* are animals such as fish and squids that swim actively and may move long distances for feeding or breeding. *Benthos* is the third main division of marine life and includes most sea-shore life. Benthic organisms crawl about on the sea-bottom, or burrow into it or grow attached to it.

Relative humidity—the ratio of water vapour in the air (as %) to the amount that would saturate it at the same temperature.

Salinity—weight of solids, in grams, contained in one kilogram of sea water. As usually defined, excludes carbonate and bicarbonate.

Specific heat—quantity of heat required to raise a unit mass through a given temperature range, referred to the heat needed to raise an equivalent mass of water through the same temperature range.

Thermistor—a semi-conductor whose resistance to an applied electric current varies proportionally to its temperature.

Thermocouple—a combination of two dissimilar metals (or semi-conductors) which produce an electric current when connected in a circuit and kept at a different temperature to the rest of the circuit (which is usually another thermocouple). The current is proportional to the temperature difference.

Thixotropy—property of some gels and of beach sands and muds, which liquefy on being subject to pressure or shaking, and solidify again on standing.

Index

ACTINIA, 17, *18*, 21, 87, 92, 110
Aeolidia papillosa, 24
Alaria, 39
  esculenta, 26, 32
algal life histories, 108
Ammodytes, 59
Amphipholis, 25
Amphitrite, 62, *66*
Anemonia sulcata, 17
Aplysia, 42
aquaria, establishment and maintenance of, 137
Arabella iricolor, 72
Archidoris, 24, 42
Arenicola, 56, 59, 61, 68, *91*, 92, 95, 125
  marina, *63*, 64, *65*, *66*
Arenicolides, 72, 96
Artemia salina, 79, 139
Ascophyllum, 19, 28, 44, 87, 115
  mackai, 64
  nodosum, 17, 39
Asterias, *23*, 24, 92, 96
Audouinia tentaculata, *63*, 72

BALANOPHYLLIA REGIA, 42
Balanus balanoides, *18*, 31, 39, *40*, 48, 104, 106
  larva, *99*, 100
  crenatus, 31, 106
  improvisus, 76, *83*, 104
  perforatus, 31, 39, *43*, 90

Bathyporeia, 55, 56, *60*, *62*
beach profile, 13, 51
Bifurcaria bifurcata, 32, *41*
Blennius pholis, 26
Bostrychia, 81
Botrylloides, 21
Botryllus, 21
  larvae, 102
  schlosseri, 25
breeding and larval development of shore organisms, 95
burrowing, of sand and mud animals, 68

CAMPANULARIA FLEXUOSA, 97
Cancer, *20*, 84
Capitella, *57*, 61
Carcinus, 61
  maenas, *20*, 61, 72, 84, *111*
  larva, *99*
Cardium, 59, 62, *65*, *66*
Caryophyllia smithi, 42
Cereus, 59
Chaetomorpha, 76
Chaetopterus, 59
chemoreceptive senses, 107
Chorda filum, 62
chromatophores, 110–113
Chthamalus, 31, 39, 106
  stellatus, *18*, 29, *40*, 94, 119
  larvae, 100

INDEX

Cirratulus, 57, 61
cirri, of barnacles, 45
Cladophora, 76, 107, 108
Clava squamata, 17, 24, 87, 97
Clibanarius, 20, 82
Clymene oerstedi, 63
Codium, 108
competition, between species, 104
Corallina, 19, 39
Convoluta, 110
Cordylophora, 77
Corophium, 61
  crassicorne, 63
  volutator, 65
Coryne muscoides, 97
Corystes, 59
Crangon, 61, 83
  larva, 99
crevice fauna, 72
cuttlefish, colours and behaviour 113
Cystoseira, 32, 34, 39

DELESSERIA, 107
Dendrodoa larvae, 102
  grossularia, 25
deposit feeders, 65
depth of animals in sand and mud, 64
desiccation, resistance to, 86
Diadumene, 77
Diogenes pugilator, 20
Donax, 68
Dynamena pumila, 17

ECHINOCARDIUM, 59, 63, 69
Echinus, 44
  eggs, 103
  larva, 102
  esculentus, 25
    larva, 98

Elminius, 103, *105*, 106
Ensis, 58, 59
Enteromorpha, 16, 48, 62, 76, 80, 86, 107, 119
estuaries, hypersaline, 80
estuarine animals, origin of, 77
estuarine life, adaptation to, 82
Eupagurus bernhardus, 20
Eurydice, 55, 63
evaporative cooling, 88
excretion, adaptations for, 94
exposure to air, 27

FEEDING, ADAPTATIONS FOR, 93
fertilisation, artificial, 102
filter-feeding, 44, 93
freezing-point, of blood, 84
fresh-water species, in estuaries and sea, 84
Fucus, 101, 106
  distichus f. *anceps*, 36, *41*
  serratus, 19, 31, 39, 115
  spiralis, 17, 28, 39, 115
  vesiculosus, 17, 28, 39, 76, 115, 116

GALATHEA SQUAMIFERA, 20
Gammarus, 77, 78, 83
Gibbula, 17, 19, 44
  cineraria, 22
  umbilicalis, 22
Gigartina, 107
glacial relicts, 85
Gnathia, 72
Gobius paganellus, 26
Golfingia minutum, 72, 79
grazing, 44

HALCAMPA, 59
Haustorius arenarius, 60

INDEX

Himanthalia, 31, 39
humidity, measurement of, 134
  relative, 87–89

INTERSTITIAL FAUNA, 69

LAMINARIA, 34, 39
  digitata, 19, *26*
  hyperborea, 21, *26*
  saccharina, 19, *26*, 62
Lanice conchilega, 59, *65*
larvae, collection of, 100
  dispersal of, 103
  gregarious settlement of, 106
  rearing of, 103
  selection of habitat by, 104
Laurencia, 19, 39
Leander, 17, 61
Lepidonotus clava, *18*
Leptasterias mulleri, 96
Leptosynapta, 58, 59
levelling, 122, *123*
Lichina confinis, 31
  pygmaea, 31
Ligia, 44
  oceanica, 39
limpets, homing response of, 118
Lithothamnion, 19, 21, 32, 39
Littorina, 19, 34
  littoralis, 17, *22*, 45, 96, 101
  littorea, *22*, 61, 96, 106, 117
    larvae, *98*
  neritoides, *22*, 31, 39, 94, 96
  saxatilis, *22*, 45, 31, 39, 96
Lomentaria, 107
Lysidice, 72

MACOMA, *60*, 62, 73
Marthasterias, *23*, 24
melanophores, 111, 112
Mercierella, 77, *79*

Microcerberus, 71
microclimate, 89, 134
Microhedyle, 71
mixed shores, 72
Monodonta, 45
Morchellium argus, 25
  larvae, 102
Mya, 59, 73
  arenaria, *58*, *65*
Mysis oculata, 78
  relicta, 78, 85
Mytilus, 82
  larvae, 102
  edulis, *18*, 32, 98
    larva, *98*

NEMALION, 107
Nephthys, 64, 68, 91, 125
  hombergi, 57
  longosetosa, 60
Nereis, 64, 92
  diversicolor, *60*, 61, *63*, 77
    larva, *95*
  pelagica, 21, *23*
Nerine cirratulus, *60*
Nucella, 17, 19, 96, 101, 119
nutrition, of animals in sand and mud, 64
  rocky shore, 44

OBELIA, 97
Ophiothrix, 25
Orchestia, 16
orientation, of *Littorina*, 117
organic content, of sand and mud, 52
organic matter, analysis of, 129
oxygen consumption, 117
oxygen content, measurement of, 136

INDEX

PALAEMONETES, 78
Paracentrotus, 42, *43*, 44
Patella, 106
　larvae, 102
　aspera, 39
　depressa, 39
　vulgata, 39, *47*, 87, 119
Patina, 21
particle analysis and measurement, 49, 127
Peachia, *58*, 59
Pecten, 62
Pelvetia, 17, 28, 39, 81
pH, measurement of, 136
photosynthesis, experiments on rate of, 116
Phyllodoce, 101
　maculata, *63*, 96
　larva, *95*
pigment cells, 110 *112*
pigments of algae, 107, 116
Pilayella, 17
plankton nets, 101
plane table survey, 124
Polysiphonia, 17, 107
Pomatoceros, 18, 21, 34, 39
Pontocrates, *60*
Porphyra, *16*, 86, 119
Portunus, 84
Potamilla, 72
Protohydra, *71*
Psammechinus miliaris, 25
Psammodrilus, 71
Pygospio elegans, *63*, *65*

RESPIRATORY EXCHANGE, ADAPTATIONS FOR, 90–93
Rhodymenia, *16*, 19
rhythms, 109–112
ripple markings, 51, Pl. VII

SABELLA, 62, *66*, *91*, 92
Sabellaria alveolata, 32, 39
Saccoglossus, 59
Saccorhiza, 19, *26*, 39
Sagartia elegans, 21
salinity gradient, 74, *75*
salinity, measurement of, 135
　regulation of, 82
salt marshes, 80
samples, preservation of, 132
sampling methods, 125–127
sandy shores, universal features of, 72
Sarsia tubulosa, 97
Scoloplos armiger, *60*, 96
Scoloplos larva, *95*
Scrobicularia, 62, *65*
sea water, artificial, 138
Spatangus, 62
Spirorbis, 19, 21, 34, 106
　borealis, *18*, 39
Sthenelais boa, 72, *79*
succession, of plants on rocky shore, 48
sulphur cycle, *53*, 54
surveying, 121
suspension feeders, 67

TALITRUS, 11, *55*, 56
Talorchestia, 56
Tellina, 62
　fabula, *63*
　tenuis, *60*, *66*
temperature adaptations, 88
temperature, measurement of, 133–134
tide, 5–9
tide reading, in field, 124
Trivia, *18*, 21

UCA, 82, *109*, *110*

Ulva, 107
Urothoe, 55, 56, *60*

VENERUPIS, 62, 73
Venus, *58*, 59, *63*
Verruca, 106
Verrucaria maura, 31, 39

WATER CONTENT, OF SAND AND MUD, 52
water loss, 87

wave action, 10, 28

ZONATION, CAUSES OF, 114–117
  in Britain, 27, *29*, *30*, 38–39
  in Mediterranean, *35*
  on rocky shores, *33*, 38–39
  sharpness of, 119–120
  universal, *33*, 34
Zostera, 62
  marina, 64
  nana, 64, 80

DATE DUE

| APR 7 '67 | | | |